Studies in the history of long-run equilibrium theory

To Samuel Hollander

Studies in the history of long-run equilibrium theory

Tom Kompas

Manchester University Press
Manchester and New York
Distributed exclusively in the USA and Canada by St. Martin's Press

Published by Manchester University Press
Oxford Road, Manchester M13 9PL, UK
and Room 400, 175 Fifth Avenue, New York, NY 10010, USA

Distributed exclusively in the USA and Canada
by St. Martin's Press, Inc., 175 Fifth Avenue, New York,
NY 10010, USA

British Library Cataloguing-in-Publication Data
A catalogue record for this book is available from the British Library

Library of Congress cataloging in publication data
Kompas, Tom.
Studies in the history of long-run equilibrium theory / Tom Kompas.
p. cm.
Includes bibliographical references (p.) and index.
1. Equilibrium (Economics) I. Title.
HB145.K66 1992
339.5—dc20 91-38688

ISBN 0 7190 3466-3 *hardback*

Typeset in Times 10/12pt
by Best-set Typesetter Ltd., Hong Kong
Printed in Great Britain
by Redwood Press Ltd, Melksham

Contents

Preface

These studies are devoted to the history of neoclassical economic thought. The emphasis is on a thorough re-examination of the theory of long-run equilibrium in the works of Walras and Wicksell, with special attention given to the *Elements of Pure Economics*, Volume I of the *Lectures on Political Economy*, and *Value, Capital and Rent*. Parallel developments in the neo-Walrasian theory of an 'intertemporal equilibrium' and Walras's celebrated *tâtonnement* are also discussed at length.

In sharp contrast to standard accounts, I argue that late nineteenth-century economic theory is not riddled with error. Both Walras and Wicksell set out clear and consistent theories of long-run equilibrium for a stationary economy, in essentially the same manner as more modern neoclassical approaches. I also show that the *tâtonnement* is totally unrelated to the issue of whether or not equilibrium is stable, and that Walras's introduction of 'tickets' in the theory of production was not designed to avoid the difficulties associated with path-dependent equilibria.

As I mention in the Introduction, each of the main chapters can be read independently. This accounts for some necessary repetition of the argument — especially with regard to Walras — throughout the book. For the reader who may find this irritating, I apologize. I also want to warn that the notes collected at the end of each chapter or appendix often greatly qualify and extend the material in the main text, perhaps more than usual for this sort of work. I urge the reader to avoid the tendency to skip over them.

In spite of the fact that these studies are basically historical, I believe that they also provide a good deal of insight into the prevailing structure of long-run neoclassical theory, in particular, to long-run systems where heterogeneity of the capital stock matters. In this regard, however, there is a part of the story that is largely missing from this book. As I see it, the neoclassical preoccupation with the structure of some given set of equations and unknowns for long-run equilibrium all but precludes a substantive analysis of the dynamics or stability of such systems. The reader will get a sense of what I am driving at in Chapter 3, in Walras's uncomfortable handling of the 'move' from Parts IV to V of the *Elements*. In a context which clearly requires the analysis of a time-sequence of prices and quantities, the best that Walras can do is to compare two different long-run systems, each already *in* equilibrium. I think this sort of problem is generic to all neoclassical theory, past and present, and I feel that it is a deeply important issue to consider. Even where an explicit dynamic process is analysed, it often turns out to be the case that such a system is so sufficiently unstable that further (often very restrictive) conditions are needed to guarantee long-run convergence, conditions that amount to little more than imposing an equilibrium at the very outset. Appendix D broaches the issue, but I pursue the matter at length, among other things, in *On the Development of Long-Run Neoclassical Theory*, soon to be published by Routledge.

One final point. A recent work by Negishi (1989) came into my hands too late to be incorporated in the text. Negishi claims, as I do, that Walras's system can be interpreted as a stationary long-run equilibrium. Unfortunately, Negishi is reluctant to attribute such a construction to Walras, opting instead for the view that the Part V system has 'to be modified' to be stationary. I believe this not to be the case, and that the nature of the *tâtonnement* for capital formation makes it clear that Walras's long-run theory is stationary, as originally constructed. In other words, Negishi is apparently content with the practice of imposing a steady-state assumption on the *Elements*, albeit now at a zero rate of growth. In this sense, his is no different than the interpretations of Walras offered by Schumpeter and Montgomery, ones which I take great exception to in this book.

Although its relative length may not suggest so, I have been working on this book for a long time, over which I have benefited greatly from discussions with the following colleagues: David Laidler and Peter Howitt of the University of Western Ontario; Geoff Harcourt (Cambridge), Robert Clower (South Carolina), and Tom Rymes (Carleton); Don Campbell, Larry Epstein, Allan Hynes, and Dušan Pokorny of the University of Toronto; Larry Levine (New Brunswick), and Brenda Spotton (York); and Tom Asimakopulos, Johanna Francis, Chris Green, and Meera Johri of McGill University. I wish to thank them all. I owe a special debt to Sam Hollander of the University of Toronto, who has kept a critical eye on virtually every line of the manuscript since its inception. With gratitude and admiration, I am pleased to dedicate this book to him.

Thanks are also due to the editorial staff at Manchester University Press, in particular, Jane Carpenter, Katharine Reeve, Juanita Bullough and Elaine White, for valuable assistance with the text and for noteworthy tolerance with my repeated delays.

T. K.

Chapter 1

Introduction

The following studies deal with the history of late nineteenth-century long-run theory. Although each study can be read independently an overall theme ties them together, namely a concern with long-run systems that define a *stationary* equilibrium as set out by Walras and Wicksell.

I begin, in Chapter 2, with a discussion of the *tâtonnement* as a way of characterizing the 'convergence' to equilibrium. Without a proper appreciation of the nature of this argument it proves impossible to comprehend Walras's theory of long-run equilibrium.

Indeed, finding a proper understanding of the *tâtonnement* is a difficult exercise, in large part because a familiarity with modern stability theory tends to distort an accurate reading of the *Elements*. *Tâtonnement* has today come to stand for a theory of multi-market stability with recontract, typically in a simple exchange context [e.g., Uzawa (1960), Negishi (1962) and Fisher (1983)]. The recontracting scheme, by design, excludes out-of-equilibrium transactions, or trades at 'false prices', to use Hicks's (1946:128) terminology.[1]

But, for Walras, things are quite different; stability of equilibrium and *tâtonnement* are separate issues. Stability is his concern only in two-good exchange, Part II of the *Elements*, where multiple equilibria are distinctly possible. *Tâtonnement*, however, first appears in Part III, multi-market exchange, where equilibrium is generally seen as both unique and stable. Multi-commodity exchange, in other words, as well as exchange with production (Part IV) and capital formation (Part V), for Walras, implies a unique equilibrium and uniqueness is seen as sufficient for stability. At the point in the *Elements* where *tâtonnement* is first introduced, the question of whether or not equilibrium is stable is no longer relevant.

Moreover, the 'convergence' to equilibrium values, contained in Part III, does *not* explicitly employ recontracting throughout. In the context of arbitrage, disequilibrium trading is carefully noted and discussed. The famous use of 'tickets' as a recontracting device, with *tâtonnement*, does not appear until Part IV, or until the theory of production.

In the *Elements*, in fact, *tâtonnement* is an argument that relates not to stability, in any strict sense of the word, but to what Walras (1954 [1900]:43–4) terms the 'determinateness' of equilibrium. Two conditions must be met: (a) the number of equations must equal the number of unknowns (excluding cases where equilibrium prices are non-positive); and (b) the actual (or 'empirical') movement of prices to clear competitive markets, in response to excess demands or economic profits and losses, is 'purely and simply a method of groping towards a solution of the equations' that define equilibrium. Practically, in other words, *tâtonnement* corresponds to real market behavior. It depicts the actual movement of competitive prices as equilibrium is established. Formally, it is an iterative process, designed to solve a simultaneous system, in terms of a number of equations and unknowns, defining equilibrium. Although the 'method of groping' characterizes out-of-equilibrium adjustment, Walras treats the 'convergence' to equilibrium as distinct from its stability; as if stability depends only on whether or not equilibrium is unique.[2]

But what of the rationale for Walras's introduction, in the fourth edition of the *Elements*, of 'tickets' in Part IV? One well-known interpretation [e.g., Jaffé (1983 [1967]) and Walker (1987)] claims that 'tickets', as a recontracting device, are designed to exclude the difficulties associated with 're-distributional effects' from out-of-equilibrium production and trades; to exclude path-dependent equilibria.[3] With disequilibrium transactions, more pointedly, the resulting equilibrium depends both on the initial data and the path on which this position is approached, i.e., on the particular process of 'convergence'. Since trades at false prices imply income or endowment effects, with changes in the pattern of demands, the final equilibrium will generally be different from that suggested by the system of static equations. The two will correspond only if prices happen to be 'correct', or at their equilibrium values to begin with.

Walras, so this claim goes, perceived this, perhaps as a result of criticisms by Bertrand and Edgeworth of the *tâtonnement*, before revising the fourth edition. Since 'tickets' only allow for production and trades at equilibrium (or when market-clearing prices are found), any complicating effects from changes in distribution are necessarily ruled out. Equilibrium, for Walras, is not only unique and stable, but path-independent as well.

There is a something wrong with this sort of interpretation, however; a problem first recognized by Patinkin (1965:534). If Walras suddenly realized that 'disequilibrium transactions would invalidate his equilibrium equations' in Part IV, thus rationalizing the use of 'tickets' to evade the issue, he should not have failed—as he did—to see the necessity for a recontracting device in Part III, where the possibility of 'endowment effects' also matters.[4]

Walker (1972:347) attempts to counter this point by arguing that 'tickets' are not needed in Part III since Walras merely assumes, throughout this section of the *Elements*, no trades at false prices, either *de facto*, drawing on his perception of actual stock market behaviour, or *ex hypothesi*, as a simplifying premise. But this is not the case. As mentioned above, Walras does allow for disequilibrium transactions (those associated with arbitrage) in the theory of exchange. There is no indication, however, of an awareness of path-dependency here. The same 'true' and unique equilibrium values are seen to be obtained regardless of whether or not out-of-equilibrium trades take place.

As I see it, the evidence clearly shows that Walras does *not* intentionally use 'tickets'—although they do have this effect—to exclude changes, as Jaffé (1983:230) terms it, in the 'distribution parameter'. Whenever Walras, in the *tâtonnement*, presupposes that 'no change in the data is allowed', he invariably and explicitly refers to holding aggregate quantities constant, not their distribution. The 'tickets' device, in this regard, is designed to separate arguments concerning the 'convergence' to equilibrium, from what Walras calls 'variations' in prices due to changes in aggregate initial data.

The difficulty, in more specific terms, is the following. In the *tâtonnement* for pure exchange (Part III), the total quantities of goods to be traded, treated as initial endowments or data, do not undergo any change as equilibrium is obtained. In Part IV, however, aggregate inputs become 'the essential data of the problem'. Tradeable (consumer) goods are now treated as variables, whose produced amounts in *long-run* equilibrium, with zero economic profits, depend on how given inputs are allocated between uses. To trace out, with actual production at 'false' long-run prices, a sequence of short-run equilibria (with realized profits and losses, and various quantities of produced and traded consumer goods), would necessarily involve, for Walras, confusing the process of *tâtonnement* in Part IV, with changes in initial endowments—relative to Part III—as an exercise in comparative-statics. The purpose of 'tickets' is to find the 'right' *long-run* amounts of consumer good output from the outset. In effect, the procedure avoids the idea of a Marshallian short-run, with equilibrium profits and losses, by characterizing, instead, adjustments around a given and invariant long-run position.

None of this has anything to do with excluding income or endowment effects. In fact, the *Correspondence* makes it clear that Walras did not become aware of a path-dependency problem from Bertrand and Edgeworth, if indeed (and this is doubtful) such a message is contained in their criticisms of the *tâtonnement*. The 'tickets' device is simply not constructed for that purpose in the first place. Although Walras clearly erred in not recognizing path-dependency, it is not, as Walker (1970:693) suggests of Walras, that the problem of disequilibrium transactions occasionally 'slipped from his mind'. It is, rather, that it was never there to begin with.

Chapter 3 is specifically concerned with Walras's (1954 [1900]) long-run theory as such. Various interpretations in the secondary literature perceive an error in the theory of long-run equilibrium contained in the *Elements*. The issue concerns an *apparent* incongruity between the equilibrium requirement of uniform rates of return across reproducible assets and the assumption of holding initial endowments of each capital good, in arbitrary amounts, fixed.

The trouble stems from Walras's special handling of two different (and often unrecognized) types of long-run systems. In Part IV of the *Elements*, only consumer goods are produced and no one saves out of income. All resources, including a number of different capital goods, are treated as 'gifts from nature': pre-existing and indestructible. Returns are input-specific as rentals to each resource owner. Long-run equilibrium occurs when production implies zero economic profits, so that the value of rental income just equals the value of output. In Part V, however, Walras allows for both consumer and capital goods to be produced, where the latter now depreciate with use. Savings is assumed, net of depreciation, so that the production of capital goods reads as flow additions to the pre-existing stock. Markets for the sale and purchase of ownership claims to resources are imagined. Economic profits are still zero. But long-run equilibrium implies uniform returns on the rental for each capital good, as a fraction of its supply price.

The perceived error appears in the 'move' from one system to another. Walras takes the endowment of each capital good, arbitrarily given in Part IV, as fixed in Part V. Although new capital goods are produced in Part V, they do not enter the economy until 'later', or in a 'period subsequent to the one under consideration'. This seems strange enough, but Walras also goes on to argue that the purchase prices of 'old' capital goods, from the pre-existing list in Part IV, are necessarily the same as those of newly produced capital goods in Part V. Rates of return, in other words, are equalized in Part V, with changes in the relative amounts of each capital good, but prices apparently remain constant throughout.

With a random collection of endowments in Part IV, held fixed, Walras mistakenly—so this argument goes—writes the long-run equations in Part V as if equilibrium is stationary; rates of return exclude a capital gain or loss term and relative prices are constant over time. More specifically, Walras assumes, in order to hold initial endowments constant, that depreciation allowances, technologically given and valued at constant prices, are 'plowed back' into each capital good, keeping it physically intact. This, however, implies that gross savings cannot be used to equalize rates of return in Part V, i.e., the allocation of depreciation allowances across different (and given) capital goods, to effect a change in their relative amounts, is not allowed. Long-run equilibrium in Part V, Walras (correctly) concludes, can only be established with positive net savings. But then the long-run equilibrium in Part V cannot (again, apparently) be stationary. Moreover, with new additions to the stock of capital, albeit entering at a later date, the prices of old versus new capital goods must vary over time, as relative quantities change.

There are two ways out of this perceived mess; two ways of interpreting Walras, most common in the secondary literature, that avoid these problems: (a) Part V as a temporary equilibrium, with non-steady-state or unbalanced growth; and (b) Part V as a steady-state equilibrium. The first [e.g., Hicks (1934), Collard (1973), Diewert (1977) and Morishima (1980)] takes the assumption of net savings, combined with static price-level expectations, as a characteristic of the Part V system throughout. The idea, in other words, is to interpret the Part V equations as a particular 'slice-in-time'; an equilibrium solution with endowments available (say) at time t, solved again, at time t + 1, with new endowments, altered by the capital goods produced at t. If prices in t + 1 are expected to remain unchanged from prices at t, the capital gains and losses usually associated with portfolio equilibrium over time can be ignored. On this reading of Walras, the equations for equalized returns, *although they explicitly describe a stationary equilibrium*, are to be interpreted in terms of an expected and temporary uniform real rate of return. Actual returns, with realized capital gains and losses, vary over a sequence of (non-steady-state growth) periods.

The second approach [e.g., Schumpeter (1954), Kuenne (1963) and Montgomery (1971)], as noted, assumes balanced growth in the move from Parts IV to V of the *Elements*. On this account, production in Part V requires that the growth rate for each new capital good be constant and proportional to the pre-existing list of capital goods in Part IV, already supposedly given—*although Walras clearly assumes a random collection of initial endowments*—in amounts that guarantee a uniform return from the outset. Accordingly, the equations for uniform returns in Part V are to be interpreted as conditions for long-run equilibrium with steady-state growth and positive net savings. Relative prices and the structure of the capital stock remain unchanged throughout the move from Parts IV to V, with no capital gains or losses.

The problem, however, is that it is impossible for either of these interpretations to be made consistent with Walras's own construction of long-run theory. The evidence I show is quite clear: Walras assumes neither static price-level expectations nor steady-state growth. In particular, service prices, i.e., the value of rentals for pre-existing inputs, determined in Part IV, and the supply prices of each capital good, determined in Part V, are not fixed—nor are they expected to remain unchanged—throughout the construction of long-run equilibrium. The prices of old and (physically identical) newly produced capital goods are indeed the same, once depreciation charges are accounted for, but this is true only after the equations for long-run equilibrium in Part V have been solved; only after, in other words, the *tâtonnement* process of adjusting rentals, prices, and the relative quantities of new capital goods, along with the amount of savings and the rate of return, has been completed. Walras does not err here.

In fact, the evidence, as I see it, shows that Walras's intention in Part V of the *Elements* is to set out the equations for a *stationary* long-run equilibrium. Relative prices remain constant over time, there are no capital gains or losses, and rates of return are uniform (in real terms, without steady-state growth or static expectations) across all capital goods, old and new. The only 'error' in Walras's exposition—rather more in the nature of an ambiguity—is a failure to explicitly state that the amount of net savings must be zero once a stationary equilibrium is obtained. Nevertheless, all of the Part V equations, as well as the more informal nature of the argument, indicate that this condition *is* part and parcel of his definition of the long-run with uniform returns, the precise meaning attached to 'an economy in normal operation which only has to maintain itself in equilibrium'.

The source of the ambiguity, what accounts for the appearance of Walras's system as being non-stationary throughout, is easy to find. The *tâtonnement*, as a way of characterizing the 'convergence' to equilibrium in Part V, requires that certain initial data remain unchanged. Since Walras begins in Part IV with fixed and arbitrary amounts of pre-existing inputs (including fixed capital goods), the only way to equalize returns—again, in the 'move' from Parts IV to V of the *Elements*—is in the context of positive net savings. Net additions to the stock of capital, in other words, are needed to generate uniform returns; but such additions, in terms of a sequence of *short-run* changes in the 'structure' of capital, cannot affect the stock of pre-existing inputs in the process of 'convergence'. (As we shall see, this also explains Walras's use of 'tickets' in the *tâtonnement* for 'capital formation', as well as the restriction that newly produced capital goods only enter the system 'later'.) Once the 'move' to Part V is completed, however, i.e., once *long-run* equilibrium is obtained in the Part V system, the economy is stationary. Rentals, prices, the final quantity of each capital good (pre-existing plus net additions), and the uniform rate of return are all known and constant. The practice of imposing steady-state growth or static price-level expectations on the *Elements* is unnecessary.

There is one additional interpretation of Walras in the secondary literature worth noting. I have in mind the works of Garegnani (1976) and Robinson (1961).[5] Rather than attributing an assumption of static expectations or steady-state growth to the *Elements*, Garegnani (1976:34) claims that Walras simply erred in his construction of long-run theory:

> Since Walras, no less than the other author's just mentioned [Wicksell, Clark, Böhm-Bawerk, and so on], attempted to determine a long-period equilibrium, he was simply incorrect in believing this to be compatible with treating each kind of 'capital good proper' as a separate factor in given supply. With a capital endowment conceived in these terms, the forces of demand and supply can only reach a *short-period equilibrium*, i.e., an equilibrium where the prices of the services of the capital goods will generally not be compatible with a uniform rate of profit on the (actual or potential) supply prices of the respective capital goods.

Robinson (1961:57), in a review of Sraffa (1960), while sharing this opinion that Walras is unable to obtain a long-run equilibrium without error, goes on to argue—evidently assuming that both aggregate inputs and their distribution among firms is arbitrarily fixed—that a Marshallian short-run equilibrium is the best that a Walrasian theory of supply and demand can provide:

> In a market economy, either there may be a tendency towards uniformity of wages and the rate of profit in different lines of production, or prices may be governed by supply and demand, but not both. Where supply and demand rule, there is no room for uniform levels of wages and the rate of profits. The Walrasian system makes sense if we interpret it in terms of an artisan economy, where each producer is committed to a particular product, so that his income depends on his output and price. Each can have a prospective rate of return on investment in his own line, but there is no mechanism to equalize profits between one line and another.

My interpretation, on the contrary, shows—subject to the ambiguity above—that Walras constructed a fully coherent and consistent stationary long-run equilibrium in Part V of the *Elements*; that the Part IV system is not a short-run equilibrium, and certainly not in the sense of Marshall; and that Walras's assumption of given endowments is, in fact, compatible with a long-run system generating uniform returns, through net accumulation, in a supply and demand context.

In Chapter 4, I turn to the long-run theory of Wicksell, with evidence drawn largely from *Value, Capital and Rent* (1954 [1893]) and the *Lectures on Political Economy* (1934 [1911]). A myth has developed in the secondary literature regarding the manner in which Wicksell closes his various long-run systems, each defining a stationary equilibrium. It claims that he takes the aggregate value of the stock of capital as a datum, in order to ensure that the number of equations equals the number of unknowns.

The specific context that I have in mind is a familiar one. A long-run equilibrium with reproducible inputs, is characterized by a structure of capital that generates a uniform rate of return, i.e., each capital good is produced in the relative amounts that guarantee a ratio of net rentals to supply prices that is the same for all. Since returns are equalized in real terms, capital gains and losses must be zero. All relative prices, quantities, and rentals are constant over time. In addition, if the system is stationary (in steady-state with a zero rate of growth), the production of new capital goods must simply cover the depreciation of the fully adjusted stock of capital, leaving its structure intact. Except for the case where capital can be expressed as a physical scalar (and where there is no 'period of production'), the equilibrium value of the aggregate stock of capital is endogenous to both the total quantities of each capital good produced and their relative prices.

Counting the number of equations and unknowns [see Section 3.1 below] for this long-run equilibrium—leaving savings behaviour or intertemporal preferences aside—gives a standard result. In general, the system will be 'short' an equation to determine the unknowns; a 'missing equation' (or a set of independent equations and unknowns) is still needed to close the system.

The typical neoclassical approach to this problem, e.g., Hicks (1946) or Hirshleifer (1967), is to close with an equation expressing an equality between savings and investment, with savings as a function of all prices and the rate of return. The savings function can either be specified in an *ad hoc* form [Hicks (1946)], or it can be derived [Hirshleifer (1967)] more formally through explicit intertemporal optimization, where an equality between the rate of time preference and the rate of return (at a zero rate of growth), simply adds an extra equation and unknown. Zero net savings, i.e., zero net additions to the stock of capital, defines the special case of a stationary equilibrium, where (again) relative prices and the aggregate value of capital, in particular, remain unchanged over time. It is Wicksell, so the mythology goes, who acts differently, who closes his systems of equations by erroneously fixing the measure of value-capital from the very outset.

Indeed, Hirshleifer (1967:194, 197) charges that Wicksell, in taking the value of aggregate capital as 'an autonomous constant rather than an exogenous variable of the system', simply confuses the idea of a constant value for capital over time, as an outcome of a stationary economy, with a perceived '*functional* invariance' for capital in the equations representing long-run equilibrium. Wicksell, in other words, mistakenly acts as if 'the hypothesis of the stationary state sufficed to make the system of equations complete' since capital can then be considered as an 'invariable magnitude'.[6]

Moreover, since an Austrian 'period of production' implies a trade-off between the utility from consumption 'today' versus consumption 'tomorrow', what Wicksell necessarily lacks, according to this view, is a theory of savings behaviour, an equation or set of equations and unknowns expressing intertemporal preferences, in order to explain why the equilibrium stock of capital, 'advanced' to labour, is neither being currently consumed nor accumulated. Without savings behaviour, Wicksell's system remains open—just as in Sraffa (1960)—and is compatible with any exogenously imposed rate of return which sets (in Austrian terms) the 'marginal product of time'. Without, in other words, some 'subjective factor' or a rate of time preference to determine the value of capital endogenously, the system remains, as Hirshleifer (1967:195) states it, 'a one-sided productivity model'.

A similar view is found in Neo-Ricardian literature. Garegnani (1976:33–4), for example, contrary to the typical neoclassical approach of Hicks and Hirshleifer outlined above, argues that Wicksell not only takes the value of capital as an exogenous variable, but that he has 'no alternative to doing so' if he is to be 'consistent with the conditions of long-period equilibrium'; an equilibrium where 'the marginal product of labour is the one resulting when the "quantity", but not the "form" of "co-operating capital" remains unchanged'. Since an equality, so this story goes, between savings and investment, as a way of closing the long-run system, is 'not available' to Wicksell, the value of capital cannot be considered as 'an unknown determined by the condition of zero net savings'.[7] Wicksell, in other words, as Petri (1978:252) and Howard (1980:20) claim, is simply lost in a 'circular reasoning', adopting the 'methodologically illegitimate' position of taking the 'magnitude of value capital' as an exogenous variable, to solve for the remaining unknowns, when it in fact depends on prices.[8]

A careful reading of Wicksell, however, gives a drastically different interpretation. It suggests, even with all the strange paraphernalia of Austrian capital theory, e.g., the 'degree of roundaboutness', an 'average period of production', the 'subsistence fund', and so on, that Wicksell is as thoroughly neoclassical as Hicks (1946) and Hirshleifer (1967) in approach. The stock of capital, in particular, is seen as endogenous to savings behaviour; the rate of return to capital must conform to the ratio of marginal utilities of present to future consumption; aggregate capital, when it cannot be expressed as a physical scalar, must be measured in terms of prices; and a stationary economy requires zero net savings, *determining* the equilibrium value of capital and leaving it intact over time.

The source of misinterpretation is obvious. At several points, especially in the formal models contained in *Value, Capital and Rent*, Wicksell does take the value of capital, *in strictly provisional terms*, as a datum.[9] The expressed desire is for analytical simplicity. Capital is treated as a prior accumulation—a homogeneous stock of corn with zero net savings assumed—in order to shift the focus toward the specific productivity criteria necessary to determine the rate of return, while temporarily ignoring the preference-side of the model. Wicksell in other words, simply follows the Walrasian practice of distinguishing, from an otherwise more general setting, between a 'theory of production' and a 'theory of accumulation'.

None of this, however, implies that Wicksell failed to see the necessity of closing his systems of long-run equations properly, or that he failed to set out all the conditions that are needed to do so. Even in the special case where capital can be expressed as a physical scalar—Wicksell is quite clear on this point—a theory of savings behaviour is still required to determine the stationary and endogenous amount of capital. The necessity for a system of equations and unknowns expressing intertemporal preferences is fully allowed.

Chapter 5 concludes with the recognition of an 'essential continuity' between the works of Walras and Wicksell. Each sets out what amounts to the same theory of a stationary long-run equilibrium, a point that Wicksell came to realize near the end of his career. I also draw, in this chapter, some general implications of my interpretation of Walras and Wicksell for the history of economic thought.

I close these studies with four appendices. Appendix A provides the relevant exegesis regarding Wicksell's recognition of an 'essential continuity' in long-run theory. In Appendix B, I detail the growth-path that results from changes in the equilibrium amount of capital. Wicksell, in particular, defines intertemporal preferences in such a manner that the only possible steady-state, in the presence of land as a fixed input, is the special case of a Classical stationary state, in which the rate of time preference, the rate of growth, and the rate of return on capital are all zero. Appendix C shows that this 'essential continuity' carries forward through Hicks and the Neo-Walrasian idea of an intertemporal equilibrium, while Appendix D examines the difficulties associated with the stability of a stationary equilibrium.

In the secondary literature, to summarize, the history of late nineteenth-century long-run theory is invariably portrayed as a history of error. What I demonstrate in the following studies, in sharp contrast to these accounts, is that both Wicksell and Walras set out a fully coherent and consistent theory of long-run equilibrium for a stationary economy. There is *no* history of error here. First, Wicksell, except for certain well-defined purposes and in strictly provisional terms, does not treat the value of capital as an exogenous variable. Second, Walras's economics is not limited to short-run theory nor does his theory of long-run equilibrium require the auxiliary assumptions of steady-state growth or static price-level expectations. Both Wicksell and Walras properly close their systems of long-run equations with a savings function and an equality between (zero) net savings and investment—essentially the same procedure used in the more modern approaches of Hicks (1946) and Hirshleifer (1967).

What I am combating in this work is the almost universal belief that one can view late nineteenth century theory, in particular, *stationary* long-run theory, solely in terms of a Clark-like construction. Clark (1965 [1900]), in taking the value of aggregate capital as a datum throughout *The Distribution of Wealth*, is not neoclassical in the sense of Wicksell and Walras. Wicksell and Walras, by leaving the value of capital endogenous to both relative prices and the savings behaviour that defines a stationary equilibrium, are as thoroughly neoclassical as Hicks and Hirshleifer.

Notes to Chapter 1

1 The adjustment mechanism stipulates that prices must move in the same direction as the value of excess demands. Gross substitutability, i.e., an increase in the price of one good increases the excess demand for all other goods, is a sufficient condition—with excess demand functions homogeneous of degree zero in relative prices, and so on—for initial prices (arbitrarily given) to converge, with recontracting, to their path-independent equilibrium values.

2 A point first noted by Patinkin (1965:539).

3 Kaldor (1960 [1934]:19) also shares this view of Walras, defining path-dependent equilibria as 'indeterminate'.

4 As a result, Patinkin (1965:534) concludes that 'Walras was not fully aware of the logical necessity for recontracting.' This view, contrary to my interpretation, leaves the rationale for 'tickets' obscure.

5 See Petri (1978) as well.

6 The belief that Wicksell treats the aggregate value of capital as a datum is a common one in the secondary literature. Note, for example, Samuelson (1982:301): 'It is well known that Wicksell's capital theory is one equation short of being determinate, and that Wicksell addresses this problem by assuming the value of capital to be given exogenously.' Lutz (1968), Negishi (1982b), Sandelin (1980) and, as I am about to note in the text, Garegnani (1976), Petri (1978) and Howard (1980), all argue the same.

7 Garegnani does not explain why Wicksell has 'no alternative' to treating the aggregate value of capital as a datum. He appears to believe, in spite of the approach used by Hicks (1946) and Hirshleifer (1967)—one, for some reason, also 'not available' to Wicksell—that neoclassical theory requires this device. I develop this point further in Chapter 5 below.

Note, however, that taking the aggregate value of capital as a datum does imply that savings behaviour cannot be used to close the system, i.e., with the value of capital fixed rather than endogenous, an equality between savings and investment and a savings function, adds two equations in only one unknown, leaving the system indeterminate.

8 Note Pasinetti (1977:29) on Wicksell: 'Whereas the wage rate and the rent per unit of land refer to *physical units* of labour and land respectively, the rate of profit is a percentage rate referred to the *value* of capital goods. And the value of total capital is a magnitude of which requires a prior knowledge of the rate of profit, which is precisely a variable that one seeks to determine.'

This criticism of Wicksell—obviously quite damaging—is clearly prior to the difficulties, in a comparative-statics context, associated with capital reswitching and reversal [e.g., Harcourt 1972]. If taking the value of aggregate capital as a datum is seen as Wicksell's method, his theory is inconsistent right from the beginning.

9 In all but one of the systems in the *Lectures on Political Economy*, including ironically, the celebrated wine model, i.e., the model used in much of the secondary literature as evidence that Wicksell takes the measure of value-capital (K) as a datum, Wicksell explicitly writes K as an endogenous variable.

Chapter 2

Walras's *tâtonnement* reconsidered

The central issue in this chapter is the rationale for the celebrated use of 'tickets' in Part IV of the *Elements*. I show, as discussed in Chapter 1 above, that 'tickets' are *not* designed by Walras to exclude path-dependent equilibria, generated through income or endowment effects as a result of trades at false prices. In fact, there is no evidence to suggest that Walras is even aware of the problem of path-dependency. The point of 'tickets', instead, is to separate arguments concerning the 'convergence' to a given (and invariant) position of equilibrium, from the 'variations' in prices due to changes in aggregate initial data.

The order of the Sections below follows that contained in the *Elements*. Specifically, 2.1 discusses Part II of the *Elements*. Here, generally speaking, Walras simply excludes trades at false prices. Since only two goods are exchanged, with individuals possessing utility over both, the principal concern is with non-unique equilibria, and hence (the only point in the *Elements*) stability. Bertrand's criticisms of Walrasian market adjustment are also considered.

Section 2.2 deals with Part III of the *Elements*. Before *tâtonnement* is first introduced, Walras discusses the effects of disequilibrium transactions associated with arbitrage. The same 'true' equilibrium prices, however, are seen to be obtained regardless of whether or not trades at false prices occur. There is no recognition of path-dependency by Walras and the *Correspondence*, including Edgeworth's criticisms of the *tâtonnement*, does not essentially deal with the issue. In addition, Newman's (1965:102–3) conjecture that the *tâtonnement* 'was meant to cope with' the complexities associated with 'convergence' in '*multiple market* situations' is shown to be correct. Newman (1960:102) errs, however, in suggesting that Walras 'explicitly allowed' for the path-dependent effects of out-of-equilibrium trades, but that the problem was 'considered minor' by Walras, not 'at all serious for *exchange*'.

Sections 2.3 and 2.4 deal with Parts IV and V of the *Elements* respectively. Not only is my interpretation of the role of 'tickets' in the *tâtonnement* consistent with the evidence in the *Elements*, it also explains, with regard to Part V (the 'Theory of Capital Formation'), in particular, a peculiar feature of Walras's theory of long-run equilibrium: in establishing uniform returns in the 'move' from the Part IV to the Part V system, Walras initially restricts the analysis to positive net savings, but also assumes that new capital goods, with accumulation, only enter the economy 'later' or in a 'period subsequent to the one under consideration'. The use of 'tickets', in this regard, is specifically designed to keep the aggregate quantity of each factor of production—relative to a given equilibrium position—constant in the process of 'convergence'.

Since, in other words, Walras begins in Part IV with arbitrarily given amounts of inputs, he has no other option but to allow for the net production of new capital goods to determine a uniform rate of return. Nevertheless, allowing these new capital goods to enter 'now'—in terms of a sequence of short-run equilibrium amounts produced—would, again, for Walras, necessarily confuse the process of 'groping' with changes in the underlying data as an exercise in comparative-statics. 'Tickets' thus avoid the problem by finding the correct relative amounts of capital goods (new plus those from the pre-existing list in Part IV) consistent with long-run uniform returns—the given equilibrium position—from the very outset, before any actual production takes place.[1]

2.1 Part II of the *Elements*

Walras builds his general equilibrium systems in the *Elements* sequentially, with increasing generality. The theory of exchange, with arbitrarily given stocks of goods to be traded, is broadened to include a theory of production where the number of goods, as produced output, can be varied. The theory of production, in turn, with given inputs, is broadened to include a theory of capital formation where the amounts of each capital good, in particular, are variable. In Part II, Walras starts with the simplest case: two-person, two-good exchange.

Part II, Lesson 5, begins with a general discussion of competitive pricing. The 'best organized' markets, for Walras (1954:84), are 'those in which purchases and sales are made by auction, through the instrumentality of stockbrokers, commercial brokers or criers acting as agents who centralize transactions in such a way that the terms of every exchange are openly announced and an opportunity is given to sellers to lower their prices and to buyers to raise their bids.' An exchange rule (in either theoretical or practical terms), in this context, guarantees that trades only occur at equilibrium prices.

The stock markets of Paris or London are taken as prime examples. 'Theoretically,' if effective demands exceed offers for 'titles', Walras (1954:85) observes, 'trading should come to a halt.' The market price announced by brokers increases. If offers exceed demands, 'trading stops' and prices fall. 'Exchange takes place' only when offers and demands are equal, implying that the 'market is in a *stationary state* or in *equilibrium*'.[2]

It is important to note that the explicit exclusion of trades at false prices in the market for titles was not made by Walras until the second edition [1889] of the *Elements*.[3] There appear to be two reasons for the new entries. In practical terms, Walras (1880), as Jaffé (1983:247n13) remarks, undertook a detailed study of the actual operation of the Paris stock exchange after the first and well before the second edition of the *Elements* appeared. The evidence made it clear that exchange only at equilibrium prices was, in fact, the operating rule of such markets. In large part, this is the way the issue is presented in Lesson 5—as a matter drawn from actual experience.

In theoretical terms, the new entries may have been prompted by Bertrand's (1883) review of an earlier work by Walras (1883). Here, Bertrand sketches an argument to the effect that out-of-equilibrium trades, in pure exchange, imply that the position of the offer and demand curves varies 'during the duration of the market', leaving the 'static' equilibrium price indeterminate. This view (perhaps) comes close to outlining the problem of path-dependent equilibria, since the order, for Bertrand, in which trades are executed—relative to the various preferences of buyers for amounts purchased at given prices—matters for the determination of the 'final' price.

Walker (1987a:766–7), indeed, uses this as evidence for Walras's introduction of the 'tickets' device in *tâtonnement*: 'The natural presupposition is that Walras eventually realized that disequilibrium transactions and production must also be eliminated from a system in which production occurs if his equations were to be valid.' Leaving aside for now, however, the fact that Bertrand's review had nothing to do with either production or 'tickets', Walras's (1936 [1898]:432) response to Bertrand is unequivocal:

> The market for titles must be a model market, and the model market, as we have said, is that wherein sales and purchases are made by the intermediary of the agents who receive the orders of sellers and buyers and operates only after having determined, in terms of present price, the price for which actual supply and demand are equal.

And, earlier in the discussion:

> In the theoretical market, if demand exceeds supply or supply exceeds demand, no one is satisfied, instead exchange remains suspended until the raising or lowering of price leads to the equality of supply and demand; after which everyone is satisfied. The theoretical equilibrium price is essentially a unique price, resulting at a given moment, in a general exchange. Under these conditions, the problem of exchange is perfectly determinate. (1936:352n)

Disequilibrium trades, in other words, are simply excluded *ex hypothesi*, with no explicit acknowledgement of a path-dependency problem. Although this clearly does not answer Bertrand's objections, it must also be noted that Bertrand's points are not well presented. Even though supply and demand curves are seen to shift with out-of-equilibrium trades, Bertrand does not provide a clear or convincing rationale (e.g., income or endowment effects) for such a result. Disequilibrium trades are simply assumed to occur—reflecting the possible effects of a new buyer, with infinite demands at a price just below equilibrium, entering the market at the 'last minute', before trades take place; the influence of some unexplained or 'accidental circumstances' which result in variations around an equilibrium price; or the impact of bilateral exchange.[4] Not surprisingly, none of this had any effect on Walras's handling of competitive arbitrage (Part III of the *Elements*), where disequilibrium trades are considered, but the problem of path-dependency (whatever its source) is not recognized.

To return to Part II of the *Elements*, Walras (1954:87–91) next sets out the equations for two-good exchange, Q_a and Q_b, indexed by j where appropriate.[5] The initial quantities of each good are taken as arbitrarily given. A broker coordinates bids, changing prices accordingly, until offers and demands for each good are equal. Walras works first not with functions, but simply in terms of ratios of offers to demands, i.e., the offer and demand for Q_a is given by

$$O_a = D_b P_b \qquad (2.1.1)$$

and

$$D_a = O_b P_b \qquad (2.1.2)$$

for P_b the relative price of Q_b in terms of P_a, where

$$O_b/D_b = D_a/O_a = \alpha \qquad (2.1.3)$$

defines the relevant ratios. For $\alpha = 1$, the 'market will be in equilibrium', or $D_a = O_a$ and $O_b = D_b$ at a corresponding relative price.

The question, of course, is what happens if $\alpha \neq 1$. Here, Walras (1954:90–1) explicitly disassociates the analysis from his earlier discussion of the market for titles. The reason, however, is not to allow trades at false prices—again, the device of a broker is used to conduct bids—but that, in two-good exchange, multiple equilibria are possible. As mentioned in Chapter 1 above, this is the only case where non-uniqueness and, as a result, for Walras, instability are a concern. The rationale is clear. Since traders 'possess direct utility' over both goods, the offer of (say) Q_b is generally not single-valued. If the demand for Q_a becomes inelastic, a fall in the relative price P_a implies the need for less of a value-offer of Q_b in return.[6] The offer of Q_b (and likewise for Q_a) thus 'starts at zero, increases, attains at least one maximum, then decreases and returns to zero'; with the possibility of an unstable equilibrium.

To handle this, in Lessons 6 and 7, Walras (1954:92–114) now considers demands and offers as the functions

$$D_j = F_j(P_b) \quad (2.1.4)$$

and

$$O_j = G_j(P_b) \quad (2.1.5)$$

where P_b, again, is the relative price of Q_b in terms of P_a. The familiar condition for a stable equilibrium is

$$F_j'(P_b) - G_j'(P_b) < 0 \quad (2.1.6)$$

for $F' = \partial F/\partial P_b$, so that an increase in P_b, in response to an excess demand for Q_b, lowers the quantity demanded and raises offers.

Note, in particular, that Walras (1954:109) gives no indication here (as well) that equilibria are path-dependent, e.g., for the stable case:

> Such an equilibrium is exactly similar to that of a suspended body of which the center of gravity lies directly beneath the point of suspension, so that if this center of gravity were displaced from the vertical line beneath the point of suspension, it would automatically return to its original position through the force of gravitation. The equilibrium is, therefore, stable.

The initial positions of the supply and demand curves, in short, do not change with price adjustment. Indeed, at one point, he asserts:

> We are now in a position to see clearly what the mechanism of market competition is. It is the practical solution, reached through a rise or fall of prices, of the same problem of exchange to which we have just given a theoretical and mathematical solution; but it must be understood that we do not have the slightest idea of substituting one solution for the other. The rapidity and reliability of the practical solution leave no room for improvement. It is a matter of daily experience that even in big markets where there are neither brokers or auctioneers, the current equilibrium price is determined within a few minutes, and considerable quantities of merchandise are exchanged at that price within half or three quarters of an hour. In fact, the theoretical solution would be absolutely impracticable in almost every case. (1954:106)

Actual market prices, in other words, are in effect always equilibrium prices, even without the benefit or device of a broker.

In Lesson 10, Walras (1954:143–9) ties the demands for goods to explicit utility maximization and analyzes the effects, in terms of a comparative-statics exercise, of a change in the initial data (i.e., the quantities of Q_a and Q_b possessed) on the equilibrium relative price.[7] Since Walras (1954:115–7) writes utility functions as additively separable across goods, the 'Theorem of Maximum Utility' reads as

$$P_b = \partial U/\partial Q_{bi} / \partial U/\partial Q_{ai} \qquad (2.1.7)$$

for each trader indexed by i. An increase in Q_b, for example, all other things equal, simply lowers the equilibrium price P_b.

2.2 Part III of the *Elements*

In Part III of the *Elements* Walras considers multi-good exchange where the initial quantities of the goods possessed are, once again, arbitrarily fixed. The idea of a broker to coordinate trades, across various markets, is explicitly removed—although prices are still 'cried'—and the *tâtonnement* process makes its first appearance.

In Lesson 11 Walras (1954:143–63) begins with three goods to be traded. To form an analog to Part II, exchange is first analyzed in terms of all possible pair-wise comparisons across markets, i.e., the individual holder of Q_a exchanges for Q_b and Q_c, the holder of Q_b for Q_a and Q_c, and so on, with no specific choice of *numéraire*.

Each individual, in other words, possesses two demand functions, given (e.g., for a holder of Q_b) by

$$D_{ab} = F_{ab}(P_{ab}, P_{cb}) \qquad (2.2.1)$$

and

$$D_{cb} = F_{cb}(P_{ab}, P_{cb}) \qquad (2.2.2)$$

for P_{ab} the price of Q_a in terms of the price of Q_b. The individual demands for Q_a and Q_c, for holders of Q_b, is

$$q_b - o_{ba} - o_{bc} = d_{ab}P_{ab} + d_{cb}P_{cb} \qquad (2.2.3)$$

or the difference between the value of the initial quantity possessed and the offer of q_b. 'Equations of exchange', summing demands for each good, add six equations of the form

$$D_{ba} = D_{ab}P_{ab} \qquad (2.2.4)$$

giving, in total, twelve equations to solve for 'the following twelve unknowns: the six prices of the three commodities each expressed in terms of the other two, and the six total quantities of the three commodities which are exchanged for one another.'

The extension to m different commodities is straightforward. Keeping the device of pair-wise comparisons generates 2m(m – 1) equations in the same number of unknowns: m(m – 1) prices and m(m – 1) total quantities exchanged.[8] The problem, however, as Walras (1954:157) remarks, of the exchange of several commodities for one another, as such, only 'appears to be solved':

Under the conditions described above, there would indeed be a certain equilibrium in the market so far as the prices of commodities taken two at a time are concerned; but that equilibrium would be an imperfect equilibrium. *We do not have* perfect *or general market equilibrium unless the price of any two commodities in terms of the other is equal to the ratio of these two commodities in terms of any third commodity.*

The m(m – 1) prices, in other words, leave room for gains from arbitrage through, for example, the 'indirect purchase' of Q_b, from the offer of Q_a, by first purchasing Q_c. The rationale for such gains comes from the fact that prices, across all markets, are not expressed in terms of a common *numéraire*.

To analyze this, Walras (1954:158) again partitions markets by pairs of commodities, giving 'm(m – 1)/2 special markets, each identified by a signboard indicating the names of the two commodities exchanged there as well as their prices or rates of exchange'. Room for arbitrage implies that

$$P_{cb} \neq P_{ca}/P_{ba} \qquad (2.2.5)$$

so that, as Walras (1954:160) explains:

No one of the holders of (A), or (B), or (C) will hesitate to resort to the expedient of substituting the indirect exchange of (A) against (C) and (C) against (B) for the direct exchange of (A) against (B). . . . This indirect exchange is called *arbitrage*. As to the gains that trading parties realize by arbitrage, they will distribute them as they please according to their various wants, by purchasing a little more of one commodity or another in order to procure the largest possible sum of total satisfactions. The condition of this maximum, it is well to point out, is that the ratios of the intensities of the last wants satisfied be equal to the real prices resulting from arbitrage operations. But we shall not go into that now, for it suffices to note at this point that the supplementary demand (entailed in arbitrage operations) is part and parcel of the principal demand.

For the case in which $P_{cb} > P_{ca}/P_{ba}$, arbitrage—with a change in the pattern of 'supplementary' demands—results in a rise in P_{ca} and an fall in P_{ba}, until $P_{cb} = P_{ca}/P_{ba}$. Just the opposite results when $P_{cb} < P_{ca}/P_{ba}$.

At the end of Lesson 11, Walras (1954:161–2) shows how the m(m − 1) prices in pair-wise markets reduce to m − 1 relative prices, with exchange values now 'cried' in terms of a common *numéraire*. The result is intuitive. Equations of exchange take the form

$$D_{ba} + D_{bc} + \ldots = D_{ab} \cdot (1/P_b) + D_{cb} \cdot (P_c/P_b) + D_{db} \cdot (P_d/P_b) + \ldots \qquad (2.2.6)$$

for $P_a \equiv 1$, representing m − 1 equations in the same number of price unknowns, with no opportunity for arbitrage.

All of this is quite significant. The recognition of arbitrage in exchange *must* entail, as Walras rightly acknowledges, the presence of out-of-equilibrium trades until the m(m − 1) prices in the pair-wise markets 'converge' to, what Walras (1954:119) calls, their 'true' equilibrium values, or the m − 1 prices consistent with a common *numéraire*. This clearly contradicts Walker (1972:347), who maintains that 'in the *Elements* and elsewhere Walras simply asserted that disequilibrium transactions do not occur in exchange markets.'[9]

Nevertheless, even though out-of-equilibrium trades are allowed in Part III, it is also clear that Walras does *not* recognize that such trades would necessarily change final exchange values, or that 'true' prices are path-dependent. Some hint of this is given in the above quote where Walras (1954:160) notes that the 'supplementary demand (entailed in arbitrage operations), is part and parcel of the principal demand' for goods. The final prices arrived at, in short, are no different than the ones that 'principal' demands would suggest.

But there is further evidence. Walras (1954:158–61), throughout this section of the *Elements*, literally defines 'true' prices—again, those m − 1 values obtained without out-of-equilibrium transactions, or in terms of a common *numéraire*—as the *same* prices consistent with the outcome of arbitrage. A similar view is expressed at the beginning of Lesson 12:

If each of the parties to the exchange is a holder of one commodity only, and if, in order to furnish an occasion for arbitrage transactions, the m(m − 1) prices of the m commodities are cried as ratios of exchange between the commodities taken two at a time without reference to the condition of general equilibrium, then maximum satisfaction will be achieved by each party when the ratios of the [marginal utilities] of the several commodities demanded to the [marginal utility] of the one commodity originally held are equal, not to the prices as they are first cried, but to the true prices arrived at by arbitrage. (1954:164)

And, finally, in Lesson 14, Walras (1954:185) states the 'Theorem of General Equilibrium', with a common *numéraire*, in the following manner:

When the market is in a state of general equilibrium, the m(m − 1) prices which govern the exchange between all possible pairs drawn from m commodities are implicitly determined by the m − 1 prices which govern the exchange between any m − 1 of these commodities and the mth.

Now, either this 'theorem' argues that 'true' prices are the same, regardless of whether or not arbitrage is allowed, or it is simply a truism. Starting, in other words, with m − 1 equilibrium prices will always guarantee, working back through the partition of m(m − 1)/2 'special markets', a comparable set of m(m − 1) pair-wise prices. But then disequilibrium transactions would necessarily be excluded by definition. The natural conclusion, particularly since Walras wants to account for arbitrage (i.e., the nature of 'supplementary demand'), is that he must have a theorem about 'true' prices in mind; and hence that he is *not* aware of a problem with income or endowment effects. There is certainly no explicit acknowledgement of path-dependency to be found here. Indeed, at the only point in Part III where Walras (1954:185) does allow for the effects of a change in distribution (Lesson 14, 'The Theorem of Equivalent Redistributions'), such changes are considered only after equilibrium prices have been obtained—only after the process of 'groping' is completed—leaving the value of each individual's initial endowment and 'true' prices unchanged.

In Lesson 12, Walras (1954:163–72) extends the system to include the case where individuals may initially possess a number of different commodities. Market excess demand funcitons, assumed to depend only on relative prices, are constructed for each good

$$F_j(P_b, P_c, P_d, \ldots) = D_j - O_j \qquad (2.2.7)$$

as the difference between total demands and offers, for $P_a \equiv 1$, giving m − 1 equations in the same number of relative prices.[10] The value of individual endowments is

$$\mathbf{y} = q_a + P_b q_b + P_c q_c + \ldots \qquad (2.2.8)$$

with a budget constraint of the form

$$(q_a - o_a) + P_b(q_b - o_b) + P_c(q_c - o_c) + \ldots$$
$$= d_a + P_b d_b + P_c d_c + \ldots \qquad (2.2.9)$$

for total purchases as the difference between the value of offers and endowments. Equilibrium is defined as the point in which all excess demands for goods are zero. In fact, this is where Walras (1954:170) first invokes the process of *tâtonnement*, as a solution to a simultaneous system characterizing the 'convergence' to equilibrium: 'We need only show that the upward and downward movements of prices solve the system of equations of offer and demand by a process of groping.'[11]

Walras (1954:170–2) proceeds one market at a time, changing each price in response to an excess demand or offer, holding all other prices constant.[12] If, for example, there is an excess demand for Q_b, i.e., if

$$F_b(P_b, \bar{P}_c, \bar{P}_d, \ldots) = D_b - O_b > 0 \qquad (2.2.10)$$

an increase in P_b lowers D_b and increases O_b, assuming that the offer of Q_b is positive monotonic, until $F_b(\cdot) = 0$. Clearly, however, a change in P_b alters the excess demands for all other goods, which, in turn, if their prices change, affects **y** and again alters the value of F_b.

Walras (1954:172) handles this (poorly) by simply assuming that such 'indirect influences' on F_j, from an increase in P_b, are 'self-compensating', i.e., that changes in P_j do not affect the excess demand for Q_b, leaving the 'position' of the demand function for Q_b unchanged as P_b increases. Since some prices rise, in other words, presumably as substitutes for Q_b, while others fall, as complements, the net effect of these purchases and offers 'cancel' each other out, so that the value of F_b (and **y**) is unaffected.[13] Performing the same operation for each good, e.g., lowering prices when $F_j < 0$, for Walras, clears all markets.[14]

There are a number of issues here. Although without the explicit rationale of excluding path-dependency, it is obvious that the *tâtonnement* described above does not allow for trades at false prices. The aggregate amounts of Q_j remain unchanged throughout; the amounts of q_j, held by each individual, do not change until equilibrium is obtained, and hence they do not appear as arguments in the functions $F_j(\cdot)$; and the realized value of **y**, due to the nature of the 'indirect' and 'self-compensating' effects, is unaffected in the course of establishing equilibrium prices.

In addition, since prices are 'cried' in terms of a common *numéraire*, there is no opportunity—just as in the case of exchange without the process of *tâtonnement*—for the disequilibrium trades associated with arbitrage. But, again, if Walras's reasoning is that the same 'true' exchange values would be obtained regardless of whether or not out-of-equilibrium trades take place, this only adds support to the view that Walras does not recognize a path-dependency problem. What Walras fails to perceive, in other words, is that even if prices are expressed in *numéraire* values, arbitrage aside, the 'endowment effects' that characterize trades at false prices will still affect the resulting pattern of equilibrium prices.[15]

Moreover, it is unlikely (contrary to standard accounts) that Walras could have obtained a recognition of path-dependency from the prevailing literature at the time.[16] Consider, for example, Edgeworth's (1925, II [1889]:311) famous criticism of the *tâtonnement*:

> [Walras] describes *a* way rather than *the* way by which equilibrium is reached. For we have no general *dynamical* theory determining the path of the economic system from any point assigned at random to a position of equilibrium. We know only the statical properties of the position; as Jevons' analogy of the lever implies. Walras's laboured description of prices set up or 'cried' in the market is calculated to divert attention from a sort of higgling which may be regarded as more fundamental than his conception, the process of *recontract*.

There is no suggestion that the 'position' of an equilibrium depends on the path that approaches it. Likewise, Jevons's (1965 [1991]:102–6) 'pendulum analogy' is concerned not with path-dependency, but the speed of adjustment to a determinate static equilibrium:

> It is much more easy to determine the point at which a pendulum will come to rest than to calculate the velocity at which it will move when displaced from that point of rest. Just so, it is a far more easy task to lay

down the conditions under which trade is completed and interchange ceases, than to attempt to ascertain at what rate trade will go on when equilibrium is not attained. (1965:94)

The 'better position,' for Edgeworth (1925, I [1904]:39), contrary to Walras's process of *tâtonnement*, and in line with Jevons's idea of the 'lever' or 'pendulum', is 'that only the position of equilibrium is knowable, not the path by which equilibrium is reached'.[17]

In *Mathematical Psychics*, the criticism of Walras turns more towards the possibility of multiple-equilibria with recontract and the effect of different or imperfectly competitive market structures:[18]

Where the field of competition is sensibly imperfect, an indefinite number of *final settlements* are possible; that in such a case *different* final settlements would be reached if the system should run down from different *initial positions* or contracts. The sort of difference which exists between Dutch and English auction, theoretically unimportant in *perfect competition*, does correspond to different results, *different final settlements* in imperfect competition. (1967 [1881]:47–8)

But there is still no suggestion that a competitive *tâtonnement*, with out-of-equilibrium trades, implies the presence of path-dependent equilibria. There is no recognition that an analysis of the endowment effects that result from such trades is necessarily required. Indeed, as Walker (1987b:161) points out, Edgeworth (1891:12), in an later work, goes so far as to argue that 'the give and take of bargaining by which price is determined, the direction the system follows to arrive at the equilibrium position, is not part of the domain of the science'.

Walras's response is predictable. He sees the process of *tâtonnement* as a valid description of the dynamics of 'convergence' (for a *perfectly competitive* economy) to equilibrium.[19] This is clear from a letter to Bortkiewicz [October 17, 1889, Jaffé (1965:927)], concerning Edgeworth's claim of the impossibility of a 'dynamics':

Edgeworth believes . . . I am engaged in absolutely useless exercises in my efforts to demonstrate that the operations of raising and lowering of prices, of the increases and decreases of the quantities of products produced, etc., on the markets are nothing other than the solution by *tâtonnement* of the equations of exchange, of production, and of capital formation.

And, again, to Bortkiewicz [February 27, 1891, Jaffé (1965:999)]:

> In the question of *tâtonnement*, for example, I take the almost universal regime of free competition in regard to exchange, that which was described by John Stuart Mill, and which consists in raising the price in the case of the quantity demanded exceeding the quantity supplied and lowering it in the case of the quantity supplied exceeding the quantity demanded, and I demonstrate that the process leads to equilibrium by establishing the equality of quantities supplied and demanded. Whereupon there is thrown at my head the market for English public debt, the system of English auctions, the system of Dutch auctions, etc., etc.

Pareto (1964 [1896–7]:24–5) evidently agreed with Walras's position: 'Walras has shown that the bargaining established with free competition is the means of solving the equations of exchange by successive attempts. Edgeworth has objected that the *tâtonnement* process is only *one* way. He is right; but the way indicated by Walras is truly that which describes most cases.'[20] None of this, however, deals with the nature of endowment effects.

The same can be said, contrary to Jaffé (1983:226n16), for Cournot (1960 [1838]). Like Walras, Cournot (1960:131) acknowledges that a change in the price of a single good affects the demands and offers of all other goods; but such an effect is seen as only a 'negligible' one, of a 'second order', which ultimately must progress with 'gradually decreasing amplitude'. Although there does seem to be some presumption in the analysis that final equilibrium prices may vary due to these 'negligible' effects, there is *no* recognition by Cournot of the presence of trades at false prices, i.e., of the effects of a change in the distribution of income, or the value of endowments, which result in path-dependent changes in the pattern of demands and prices. In any event, even if there is such a recognition, Walras, as Jaffé remarks, 'seems to have missed' it altogether.[21]

Note, too, as mentioned in Chapter 1 above, that Walras completely separates the process of *tâtonnement* (Part III) from the question of whether or not equilibrium is stable (Part II). Multi-commodity exchange, for Walras, invariably results in a unique and thus stable equilibrium. So strong is this perception that in Lesson 15, on 'Purchase and Sales Curves', Walras (1954:192–200) erroneously graphs all demand and offer functions as single-valued;[22] and, in the 'Geometrical Theory of the

Determination of Prices', the first appendix to the *Elements*, although all demand and supply functions, including the amount of savings, are correctly graphed as multi-valued, Walras (1954: 468–70) always picks the case of a stable equilibrium to make a point, acting *as if* it were unique. There is simply no connection between 'groping' and stability in the *Elements*.[23]

There is one final issue. Newman (1965:102–3) argues that the *tâtonnement* is a device that 'was meant to cope' with the:

> problem of convergence of the 'excess demand' mechanism in *multiple market* situations, where inter-relations of substitutability and complementarity make the analysis of convergence in simple two-commodity markets just too simple to apply without considerable change.

Jaffé (1983:237) counters with the claim that 'surely this cannot be so', since Walras writes utility as the 'sum of independent utility functions pertaining to each good separately'. In Part II of the *Elements*, Walras (1954:115–31) clearly does consider individual utility functions as additively separable across goods. In a multi-market context this implies, with a fall in the price of a particular commodity, that the prices of all other goods either rise or fall together, depending on the elasticity of demand for the particular good in question.[24] All other prices, in short, move in the *same* direction. The 'self-compensating' moves, in other words, as Jaffé (1983:237–8) observes, that Walras relies on as part of the intuition for 'convergence' to equilibrium are simply incompatible with the form of the utility functions that are assumed.[25]

The evidence, however, is unambiguous here. Newman is correct. The *tâtonnement* in the *Elements* first appears in Part III, where the case of more than two commodities is considered. The context is always one in which 'feedback' effects across various markets matter:[26]

> It is quite true that in determining the price of (C), we may destroy the equilibrium with respect to (B), that in determining the price of (D), we may destroy the equilibrium both with respect to (B) and with respect to (C), and so forth. But since the determination of the prices of (C), (D) . . . will, on the whole, entail certain compensating effects on the relationship between the demand and offer of (B), in all probability equilibrium will be approximated more and more closely at each successive step in the groping process. (1954:470)

And, in the *tâtonnement* for the theory of production:[27]

> The consequences of a change in the quantity manufactured of (B) applies also to the consequences of the manufacture of (C), (D) . . . Hence, we may be sure that the (primary) change in the quantity manufactured of any given product has a direct effect on the selling price of that product, which is all in one direction, whereas (the consequent) changes in the quantities manufactured of other products, if we suppose all those changes to take place in the same direction, will only have indirect effects on the selling price of the first product, some in one direction and some in another, thus cancelling each other out, more or less. The (resulting) system of new quantities of manufactured products and of new selling prices is thus closer to equilibrium than the old one; and we have only to continue the process of groping to approach still more closely to equilibrium. (1954:246)

As Walras sees it, the 'convergence' to equilibrium requires, given a change in the price of a single good, say Q_b, that the prices of all other goods move in *different* directions. An increase, in other words, in the price of substitutes, and a fall in complements, must leave the excess demand for Q_b, in terms of 'indirect' effects, unchanged as P_b rises. None of this can be obtained, however, with additively separable utility functions; Walras errs. But this certainly does not imply, as Jaffé suggests, that the *tâtonnement* process is not, by Walras's design, meant to deal with the complexities of multi-good exchange. Walras simply does not appreciate the technical nature of the utility functions that are necessary to make the argument work in this way.

2.3 **Part IV of the *Elements***

In Part IV of the *Elements* Walras extends the theory of exchange to accommodate the production of tradeable goods in long-run equilibrium. The context is one of long-run cost price where, for Walras (1954:224–5), the 'selling prices of products equal the costs of the productive services that enter into them', so that 'entrepreneurs make neither profit nor loss'. All resources, including a number of different kinds of capital goods, are treated as 'gifts from nature': pre-existing and indestructible. Only consumer goods are produced, establishing rentals or service prices for each input, and no one saves out of income.

Let ($\mathbf{l}$, $\mathbf{n}$, $\mathbf{k}$) define (L_1, L_2, . . . , N_1, N_2, . . . , K_1, K_2, . . .) as the given amounts of each type of land, labour and capital good, or n different services in total, indexed by i where appropriate. Walras designates Q_a, Q_b, Q_c, . . . as the output of m consumer goods, indexed by j, with prices P_j and input-specific rentals (R_l, R_n, R_k) for each service.

Individuals are assumed, in Lesson 20, to possess, once again, additively separable utility functions over all goods and services, with the total offer of each service

$$O_i = F_i(P_j, R_l, R_n, R_k) \qquad (2.3.1)$$

and the demand for each good

$$D_j = D_j(P_j, R_l, R_n, R_k) \qquad (2.3.2)$$

as functions of all prices. Individual budget constraints are given by

$$o_l R_l + o_n R_n + o_k R_k = d_a P_a + d_b P_b + d_c P_c + \ldots \qquad (2.3.3)$$

for $P_a \equiv 1$ as *numéraire*. Until Part VII of the *Elements*, per-unit input coefficients (a_{lj}, a_{nj}, a_{kj}) are fixed and given. For each (**l**, **n**, **k**), long-run equilibrium implies n equations of the form

$$O_k = a_{ka} D_a + a_{kb} D_b + a_{kc} D_c + \ldots \qquad (2.3.4)$$

so that 'the *quantities of productive services used are equal to the quantities effectively offered*'; and

$$P_j = a_{lj} R_l + a_{nj} R_n + a_{kj} R_k \qquad (2.3.5)$$

or m equations expressing equilibrium cost prices.

The system (2.3.1), (2.3.2), (2.3.4) and (2.3.5) is closed, representing 2m + 2n − 1 equations in the same number of unknowns: (a) n total quantities of services offered, i.e., as an expression of the demand for inputs to produce output; (b) the n prices of these services; (c) the m total quantities of each consumer good demanded; and (d) the m − 1 prices of these goods in terms of the price of Q_a.

It is at this point in the *Elements* that Walras (1954:242) introduces the famous device of 'tickets' for the *tâtonnement* in production, noting a qualification to the theory of exchange:[28]

We propose to establish equilibrium in production *ab ovo*, just as we established equilibrium in exchange, that is to say, by assuming the data of the problem to be any data whatsoever and yet invariable over a certain period of time, except that we may later suppose them to change in order to study the effects of such changes. But the process of groping in production entails a complication which was not present in the case of exchange.

In exchange, (the total existing quantities of) commodities do not undergo any change. When a price is cried, and the effective demand and offer corresponding to this price are not equal, another price is cried for which there is another effective demand and offer. In production, productive services are transformed into products. After certain prices for services have been cried and certain quantities of products have been manufactured, if these prices and quantities are not the equilibrium prices and quantities, it will be necessary not only to cry new prices but also to manufacture revised quantities of products. In order to work out as rigorous a description of the process of groping (towards equilibrium) in production as we did in exchange and yet take this additional circumstance into account, we have only to imagine, on the one hand, that entrepreneurs use *tickets* to represent the successive quantities of *products* which are first determined at random and then increased or decreased according as there is an excess of selling price over cost of production or vice versa, until selling price and cost are equal; and, on the other hand, the landlords, workers and capitalists also use *tickets* to represent the successive quantities of services (which they offer) at prices first cried at random and then raised or lowered according as there is an excess of demand over offer or vice versa, until the two become equal.

Equilibrium is established, in short, through the issue of 'provisional contracts' or 'pledges', for the quantity of goods and services demanded and supplied, until price adjustments guarantee a long-run outcome.[29] Production and exchange, in other words, 'effectively' take place only when '*current equilibrium prices*' are obtained. The *tâtonnement*, by design, leaves (**l**, **n**, **k**) unchanged; '*no change in the data is allowed*'.

Walras (1954:243), in fact, begins the analysis, in Lesson 21, *in* long-run equilibrium, with zero economic profits, and then sets out the effects of adjustments in Q_j, to their equilibrium values, on all prices, quantities and offers. Entrepreneurs responding to 'ticket' amounts of economic profits and losses drive the result. A Q_b, for example, which is less than its equilibrium value Q_b^*, implies that (a) the selling price P_b exceeds its cost price P_b^*; (b) the demand for services to produce Q_b is greater than their current offer; (c) the consumer demand for Q_b exceeds the quantity demanded at Q_b^*; and (d) the current rental for each service, say R_l, is less than its equilibrium rate or R_l^*.

As the output of Q_b increases, for Walras (1954:243–54), its price falls, as a move along a given demand curve, costs of production rise to P_b^*, as a move along a supply function, and rentals for each service increase until offers and demands for inputs are equal with zero profits.[30] Individual income

$$\mathbf{y} = o_l R_l^* + o_n R_n^* + o_k R_k^* = d_a + d_b P_b + d_c P_c + \ldots \quad (2.3.6)$$

at equilibrium service prices—as a formal analog to Part II of the *Elements*, where the value of initial endowments remains unchanged in the *tâtonnement* to exchange equilibrium—is assumed to remain constant throughout.

With regard to 'second order' or 'indirect' effects, Walras employs the same argument for the 'convergence' to equilibrium as that contained in Part II; the same argument, in short, applies whether or not the 'tickets' device is used. Although an increase in P_b clearly affects the value of prices and demands, and therefore service rentals and incomes, in other markets, these effects are seen to largely 'cancel each other out'. Prices and rentals, and hence consumer expenditures and incomes, rise with an increase in the output of substitutes to Q_b, and fall with complements, so that total incomes and purchases remain unchanged. The increase in P_b simply affects the excess demand for Q_b and performing the same operation on all other goods, for Walras, clears all markets at cost prices.[31]

The key question, however, concerns Walras's specific rationale for the introduction of 'tickets' in the theory of production (Part IV of the *Elements*) and not in exchange (Part III), especially since he uses the same 'groping' process—effectively excluding trades at false prices and leaving the value of **y** constant—in both theories.

Although admitting that 'there is no direct evidence of any awareness of the distributional effects of the process of *tâtonnement*' in the *Elements*, Jaffé (1983:231), nonetheless, claims that Walras introduces 'tickets' to avoid 'the "problem of the path"'.[32] To be precise, the argument runs in terms of a 'snag' that Jaffé suggests Walras 'perceived' in revising the fourth edition:

> Unless all prices of services and products were *the* equilibrium prices *ab initio*, entrepreneurs would produce non-equilibrium quantities of products at successive stages of the groping process and that these 'false quantities' once produced would have the effect of changing the parameters which he insisted (presumably in order that the system converges

to *the* equilibrium) must be held constant. In the fourth edition, Walras did not specify what parameters would be affected by the production of 'false quantities'. It could not have been the 'resources' parameter, or the technology parameter, or indeed the 'tastes' parameter (that is, the utility functions). What would, in fact, change would be the 'distribution' parameter, that is, the distribution of the asset values of the original endowments held by the individual participants in the system. (1983:230)

The device of 'tickets', Jaffé (1983:231) concludes, avoids the problem of path-dependency; without 'tickets', in other words, to exclude endowment effects, given that 'each participant's demands for products and offer of services are functions of his income as well as of market prices, whatever equilibrium the market converges to—if it converges at all—must be different from the equilibrium that would result if there were no change in the values of incomes or assets of the traders.'

The problem, however, is that Walras does not specify, in very precise terms, any reasons—Jaffé (1983:234) adds that 'it is doubtful that he ever learned to work out these reasons'—for the added 'complication' with 'groping' in production, that is 'not present in the case of exchange'. What Jaffé (1983:232) does find, quoting from the second edition of the *Elements*, is that Walras was aware of the possibility that trades at 'false quantities' would imply that some firms, producing the 'wrong' Q_j, may fail in the 'convergence' to equilibrium; a recognition, for Jaffé, that entails an awareness of the (unwanted) effects on the distribution parameter:

The groping process will be completed when, in exchange for the products they have manufactured, the entrepreneurs obtained from the landowners, workers and capitalists precisely those quantities of land-services, labour-services and capital-services which they owe to the landowners, workers and capitalists and which they have used in the manufacture of the products, so that (in the end) either they will repay their debts and shut up shop, or, more likely, they will continue their production indefinitely at the rate to which it has now become adjusted, so long as there is no change in the data, that is to say, in quantities possessed of the services and in the utilities of the products.

But this seems to have little (if anything) to do with path-dependency. And, in fact, as Jaffé (1983:233) remarks:

After calling attention [in the second edition] to the 'complication' occasioned by the production of 'false quantities', Walras immediately proceeded to ignore it in his model as if it sufficed to assume 'the data' constant to get rid of the untoward effects of the 'complication' on the endowments parameter.

Perhaps, however, this is not surprising since Walras indeed sees firms producing 'indefinitely' at equilibrium prices and quantities, as opposed to those that fail, as the 'more likely' case; a point, in any event, that seems totally unrelated to problems with the 'endowments parameter' in the *Elements*. Jaffé's argument simply presupposes that Walras's intention is to exclude 'distributional effects' in the first place.

Walker (1987a:766) also shares Jaffé's opinion of the role of 'tickets', asserting that 'Walras knew that disequilibrium transactions would invalidate his equilibrium equations.' As a first piece of evidence, Walker suggests that Bertrand 'pointed this out' to Walras in 1883. But, as discussed in Section 2.1 above, this appears highly improbable. First, it is unclear that Bertrand has the problem of path-dependency in mind; and if he does, the argument is poorly stated.[33] In any event, it is apparent from the *Correspondence*, that Walras fails to receive any such message from Bertrand (or from Edgeworth). Second, Bertrand's criticisms relate only to the exchange model. If Walras became aware of a path-dependency problem, logically, as Patinkin (1965:534) observes, he should not have neglected to introduce 'tickets' in Part III of the *Elements*, the theory of exchange—where endowment effects also matter.[34]

As a 'stronger' piece of evidence, Walker (1987:767) quotes [see p. 31 above] Walras's indication of a 'complication' in the *tâtonnement* for production. This passage, for Walker, has Walras 'stating' that 'tickets' or 'pledges' are used to exclude 'disequilibrium transactions' in order to prevent 'any change in asset holdings until the equilibrium price is found'; that 'disequilibrium production changes "the given conditions" of the production model and must be eliminated with pledges.' While it is, of course, true that 'tickets' have this effect, it is quite a different thing to see this passage as a confirmation that Walras uses this device because of an awareness of a path-dependency problem. It simply is not there. Even Jaffé, who is sympathetic to Walker's point of view, finds 'no direct evidence' for such a conclusion.[35]

Finally, Walker (1987:767–8) argues that Walras employs 'tickets' to prevent a redistribution of assets from disequilibrium trades—which necessarily implies a change in the underlying distribution parameter—in order to avoid analyzing what Walras terms a '*variable*' or '*moving*' equilibrium. The relevant passage comes from Part VI of the *Elements*, on the 'Theory of Circulation and Money':

> We shall describe equilibrium in principle, as before, first theoretically and mathematically, and then practically as it manifests itself in the market. Our economy will then be ready to pass from the static to the *dynamic* point of view. In order to make this transition we need only suppose the data of the problem, viz. the quantities possessed, the utility or want curves, etc., to vary as a function of *time*. The *fixed* equilibrium will then be transformed into a *variable* or *moving* equilibrium, which re-establishes itself automatically as soon as it is disturbed. In the theory of bimetallism we find an equilibrium of this kind. (1954:318)

All this indicates, however, is that *tâtonnement* (and note that there is no specific reference to *tâtonnement* here), for Walras, is designed to study the 'convergence' to equilibrium, holding specific quantities constant. It says nothing about a change in these amounts (used to define a unique and static position of equilibrium), as a result of income or endowment effects.

As I see it, this last point reveals the true role of 'tickets'. Throughout the *Elements*—and in all editions—Walras's chief concern in the *tâtonnement* process is to study the 'convergence' to equilibrium, holding certain data constant. In Part III (as well as Part II), 'tastes' and the aggregate quantities of Q_j to be traded are fixed and given. In Part IV it is 'tastes' and the aggregate amounts of 'original' inputs or (**l**, **n**, **k**) that are constant. (In the second edition, specifically, 'the quantities possessed of the services and in the utilities of the products', and in the fourth edition, simply 'the data'). In Part IV, in other words, aggregate inputs replace the aggregate goods to be traded as 'the essential data of our problem'.

The difficulty that Walras faces is the following. *Tâtonnement* in the exchange model clearly leaves the amounts of Q_j unchanged. In production, however, these are variables whose amounts depend on how the given inputs are allocated between uses. But the equilibrium in Part IV, by construction, is a long-run position. Allowing for trades and output at 'false quantities' must, instead, define a short-run equilibrium, with realized economic profits and losses. For Walras to proceed in this way would necessarily imply tracing out a sequence of short-run positions with consequent changes in the amounts of Q_j produced; giving, as he sees it, not an exercise in *tâtonnement*, but one in comparative-statics, or the 'variations' in prices as a result of changes in the data. Throughout the *Elements*, Walras rigorously separates the 'convergence' to a given position of equilibrium, from changes in the data which generate a comparison of equilibria.[36] The problem with the theory of production, in this sense, is that it mixes the 'groping' process with the effects of changes in the data (relative to Part III) as a result of varying amounts of Q_j.

The point of 'tickets' is to avoid this comparative-statics exercise by finding, from the outset, without actual production taking place, the 'right' mix of Q_j consistent with zero economic profits; in effect, skipping over the Marshallian short-run by characterizing, instead, out-of-equilibrium 'movements' about or in terms of the system of static equations [(2.3.1), (2.3.2), (2.3.4) and (2.3.5)] that define a given long-run equilibrium.[37]

In particular, recall that in Lesson 21 (as discussed above), Walras begins the analysis *in* long-run equilibrium, and then analyzes adjustments in Q_j, in terms of the *tâtonnement*, around this position. The idea, in other words, is to create a formal analog to the process of 'groping' in Part III by first determining the 'right' amounts of Q_j to be produced—known before *tâtonnement* begins in Part IV—and traded, along the lines set out by the theory of exchange.[38] The 'tickets' device allows this. It effectively treats the variables Q_j as if their *equilibrium amounts* can now be held constant in the 'groping' process, i.e., as if, just as in Part III, they can be listed among the data.

(This, to me, seems to be the correct meaning to attach to Walras's (1954:242) 'complication' in the theory of production. In exchange, the 'total existing quantities' of goods 'do not undergo any change' in the process of 'groping'. In production, they do. After certain 'prices for services have been cried' and certain quantities produced, if these 'are not equilibrium prices and quantities, it will be necessary not only to cry new prices but also to manufacture revised quantities of products'. In 'order to work out as rigorous a description of the process of groping' to equilibrium in production 'as we did in exchange'—constructing a formal analog to Part III—the 'tickets' device allows the 'correct' amounts of Q_j, which imply zero economic profits, to be found before actual production takes place.)

Such a role for 'tickets' appears all the more reasonable when it is remembered that Walras gives *no* evidence of recognizing a path-dependency problem anywhere else in the *Elements*: no recognition of changes in the distribution of assets changing demands; no recognition that endowment effects still matter even if prices are 'cried' in a common *numéraire*; no recognition of the need for 'tickets' in the theory of exhange (Part III) to exclude the effect of disequilibrium transactions on the 'distribution parameter'; and no response, in terms of this effect, to Bertrand and Edgeworth. Whenever Walras, in the process of 'groping', presupposes that 'no change in the data is allowed', he always and only refers to holding aggregate quantities constant, not their distribution.[39]

Now it is true, with trades in terms of a chosen *numéraire*, that Walras leaves the value of **y** unchanged in the 'convergence' to equilibrium; in effect, excluding re-distributional effects. *But the argument here, as we have seen, is the same regardless of whether or not 'tickets' are used*. The result stems not from the existence of 'pledges' *per se*, but from the manner in which Walras handles 'second order' or 'indirect' effects. Since these 'cancel each other out', the value of income remains constant in the *tâtonnement*. But this has nothing to do with path-dependent equilibria.

Note, as well, that if Walras sees the 'true' prices in Part III as the same equilibrium values that would result whether or not out-of-equilibrium or arbitrage trades are allowed [see Section 2.2 above], it is not unreasonable to conjecture that he has the same idea in mind for Part IV. Trades at 'false quantities', in other words, would presumably still give the same 'true' equilibrium values as the system converges to cost prices.[40] (The issue does not arise in Part IV, of course, since after the discussion of arbitrage all prices are 'cried' in *numéraire* values.) But, once again, if this is the case it only adds support to the view that Walras does not recognize a path-dependency problem. 'Tickets' would not have been introduced to avoid changes in the 'distribution parameter', since 'true' prices would be the same with or without disequilibrium production.

2.4 Part V of the *Elements*

In Part V of the *Elements* Walras extends the idea of long-run equilibrium to incorporate a uniform rate of return on capital goods as reproducible assets, as opposed to input-specific rentals in Part IV. The stock of each $\bar{K}_i$, however, from the pre-existing list in Part IV remains initially unchanged, with (the now assumed) depreciation allowances simply re-invested in each to keep it intact. Positive net savings imply both consumer and capital good output, with the production of new capital goods K_i, physically identical to 'old' ones—effectively changing the relative amounts of each K_i, to generate equilibrium cost prices for each capital good and a uniform rate of return. To leave (**l**, **n**, **k**) unchanged, in the 'move' from the Part IV to the Part V system, each K_i enters the economy only 'later' in a 'period subsequent to the one under consideration.'

The exercise involves adding new equations and unknowns to the Part IV system. Walras (1954:267–7) begins in Lesson 23 with the defining characteristic of long-run equilibrium (where a distinction is drawn between produced and 'original' factors), i.e., a uniform rate of return

$$r = (R_i - \delta P_i)/P_i = R_i/P_i \qquad (2.4.1)$$

on each capital good, for δ the rate of depreciation (assumed to be the same for each K_i) and R_i the net rental for each capital good. As mentioned, new capital goods, for Walras (1954:271), are produced at cost prices, adding i equations of the form

$$P_i = a_{li}R_l + a_{ni}R_n + a_{ki}R_k \qquad (2.4.2)$$

for given and arbitrary amounts of (**l**, **n**, **k**). The response to economic profits and losses again drives the result.

Individual savings is defined

$$\mathbf{s} = o_lR_l + o_nR_n + o_kR_k - [d_a + d_bP_b + d_cP_c + \ldots] \qquad (2.4.3)$$

as the difference between income and consumption, where

$$\mathbf{s} > \delta[P_1k_1 + P_2k_2 + P_3k_3 + \ldots] \qquad (2.4.4)$$

for k_i the individual holdings of each K_i, implies net accumulation, or an amount 'greater than the sum needed to cover the depreciation' of the 'existing capital goods proper' from Part IV.[41] Summing over individuals gives aggregate net savings, initially restricted to be positive, as a function

$$S = S(P_j, R_l, R_n, R_k, P_i, r) \tag{2.4.5}$$

of all prices and r. Capital market equilibrium, finally, requires

$$S = P_1K_1 + P_2K_2 + P_3K_3 + \ldots \tag{2.4.6}$$

or an equality between net savings and the value of net investment.

The system is complete. Equations (2.4.1), (2.4.2), (2.4.5) and (2.4.6) add 2i + 2 equations in the same number of unknowns: K_i or the 'quantities produced of new capital goods'; their relative prices in terms of the *numéraire*; the amount of net savings; and the uniform 'rate of net income' or r.[42]

Once again, the 'tickets' device is used to analyze adjustments around a particular long-run position; in this case, one with uniform returns, as an equilibrium obtained '*ab ovo*' or with any given amounts of 'original' inputs. As mentioned, new capital goods, for Walras (1954:282–3) only enter the system 'later':

> Thus equilibrium in capital formation will first be established *in principle*. Then it will be established *effectively* by the reciprocal exchange between savings to be accumulated and new capital goods to be supplied *within a given period of time*, during which *no change in the data is allowed*. Although the economy is becoming *progressive*, it remains (for the time being) *static* because of the fact that new capital goods play no part in the economy until later in a period subsequent to the one under consideration.

The arbitrary amounts of (**l**, **n**, **k**), in short, drawn from Part IV of the *Elements*, used now to produce both Q_j and K_i, remain unchanged in the *tâtonnement* to determine r. All of this is consistent with the static process of 'groping' contained in Parts III and IV. Aggregate quantities and tastes are held constant, with no recognition of distributional or income effects.

Walras (1954:284–9) begins in Lesson 25, in this regard, with 'random' service prices and quantities. In general, the resulting amount of S generates both

$$S \neq P_1K_1 + P_2K_2 + P_3K_3 + \ldots \tag{2.4.7}$$

or an inequality between savings and investment and

$$P_i \neq R_i/(r + \delta) \tag{2.4.8}$$

or prices for each capital good which are inconsistent with uniform returns. The value of S, from equation (2.4.5), is assumed [Walras (1954:287–8)] to be increasing in r and each service price.

Walras proceeds with equation (2.4.7) first, assuming an equality in (2.4.8), so that

$$S < [R_1/(r + \delta)]K_1 + [R_2/(r + \delta)]K_2 + \ldots \tag{2.4.9}$$

or the value of savings, for example, is less than the value of net additions to the stock of capital. (The right-hand side of equation (2.4.9) expresses the price of each capital good.) An equality in (2.4.9) is simply established through an increase in r, decreasing the 'ticket' amounts of the right-hand side and increasing the left.[43]

Next, Walras (1954:290–4) considers (2.4.8), allowing the output of each K_i to vary, assuming an equality between savings and investment in equation (2.4.7). Just as in the *tâtonnement* for Part IV, the selling price for each capital good

$$P_i = R_i/(r + \delta) \tag{2.4.10}$$

is distinguished from its cost price or P_i^*. The response to economic profits and losses drives the equilibrium result. Consider, for example, the case where

$$R_i/(r + \delta) > a_{li}R_l + a_{ni}R_n + a_{ki}R_k \tag{2.4.11}$$

or $P_i > P_i^*$, generating economic profits. Costs of production are assumed to vary inversely with the output of each K_i; own-service prices, e.g., R_i on K_i, and hence selling prices, are assumed (from Lesson 22 of the *Elements*) to fall with an increase in the output of each capital good.

Economic profits in equation (2.4.11), in other words, for Walras (1954:291–2), imply an increase in (R_l, R_n, R_k) and a fall in R_i as K_i increases, until $P_i = P_i^*$. For capital goods, on the other hand, produced with economic losses, i.e., those with relatively low rates of return, outputs and costs of production fall, while own-service prices rise. The relative amounts of each K_i change until a uniform long-run rate of return is earned on all capital goods.[44]

My view of the nature of 'tickets' in the *tâtonnement* provides a solution to a puzzle often discussed in the secondary literature: why Walras generates a long-run equilibrium by assuming positive net savings at the outset, while at the same time asserting that new capital goods only enter in a 'period subsequent to the one under consideration'.[45] The explanation is straightforward.

As we have seen, the idea of 'groping' is to describe the process of 'convergence' to equilibrium, while leaving aggregate initial data unchanged. Since Walras begins in Part IV with the fixed amounts (**l**, **n**, **k**), new capital goods, although produced, cannot by design affect the pre-existing stock of $\bar{K}_i$. The usual process, in other words, of depreciating the Part IV stock of $\bar{K}_i$ and re-investing these 'funds' (with gross savings alone) in a capital structure consistent with long-run equilibrium, would necessarily change the relative amounts of each $\bar{K}_i$. By the requirements of *tâtonnement*, Walras has no other option but to allow for the net production of capital goods to determine r.

In addition, to trace out the effects of new capital goods envisaged as a sequence of short-run (actual) additions to the capital stock would imply for Walras—just as in the various amounts of Q_j produced (with economic profits or losses) in Part IV—an exercise in comparative-statics; 'variations' due to changes in the underlying data. The point of 'tickets' is to avoid this; to avoid considering a sequence of short-run equilibria where rates of return over each K_i vary with the quantities of K_i produced. This is why new capital goods only enter the economy 'later'. The proper treatment of 'convergence', as Walras sees it, must start with the correct or *long-run* equilibrium 'pledge' amounts of K_i first.

In fact, in the *tâtonnement* for Part V (i.e., in terms of the equations and unknowns for an economy that begins *in* long-run equilibrium), Walras clearly treats all new capital goods as if they are already part of the system—just as with produced (and to be traded) amounts of Q_j^* in Part IV. The equilibrium values of R_i, P_i, K_i and r, in other words, are determined, and equations (2.4.1) and (2.4.2) imply that prices are constant over time, as if the accumulation process has already worked itself out. By the nature of the argument, Walras cannot formally begin with K_i listed among the initial data, since new capital goods are unknowns. He does not formally begin—so strong is the idea that the *tâtonnement* must leave the total inputs from Part IV unchanged—by analyzing displacements around a long-run position, with K_i (old plus new additions) endogenous and zero net savings throughout, the way Clark (1965 [1899]) does. But 'tickets' do allow Walras to treat the equilibrium amounts of K_i as if they are constant [e.g., in the *tâtonnement* for equation (2.4.9)] in the process of 'groping'.

Note, finally, that there is also no evidence for an awareness of path-dependency here. Consider, for example, the required adjustment in equation (2.4.11). Although R_i falls with an increase in K_i, with 'self-compensating' effects on all other rentals and service prices, Walras ignores altogether the 'feedback' from a change in service prices to the amount of savings, the value of income, changes in the pattern of demands, and so on.[46] Even if one excludes trades at false prices, such 'feedbacks' are surely relevant; they would not have escaped Walras had he had in mind the nature of income or endowment effects.

2.5 Conclusion

Walras (1954:83–4) begins with the recognition that the 'markets which are best organized from the competitive standpoint are those in which purchases and sales are made by auction', with brokers 'acting as agents who centralize transactions', allowing trades only at equilibrium prices. The stock market is the prime example. But a similar mechanism, even without the benefit of a broker, also applies to a wide variety of goods: 'commercial markets, grain markets, fish markets, etc.'; the markets for fruit, vegetables, poultry and groceries; doctor's and lawyer's fees, the rate for 'musician's and singer's recitals', and so on. Walras (1954:106), at one point, even suggests that actual markets clear continuously, or at least with 'the rapidity and reliability that leave no room for improvement'.

With all this in mind, it is perhaps too much to expect a recognition from Walras of the re-distributional effects contained in the *tâtonnement* process. He is aware, however, that out-of-equilibrium trades (those associated with arbitrage) do occur in exchange markets. But there is no acknowledgement that the resulting position of equilibrium will depend on the particular path on which the system converges to its stationary values.

Equilibrium is seen, outside of the context of two-good exchange, as both unique and stable. The values of the initial data, held constant in 'groping', simply help define the position of static equilibrium; a position invariant to the process of 'convergence'.

It is clear that Walras is unlikely to have obtained an awareness of path-dependency from his contemporaries: Bertrand, Bortkiewicz, Cournot, and Edgeworth. The *Correspondence* confirms that he did not. The 'tickets' device is designed not to exclude changes in the 'distribution parameter', but to leave aggregate data unchanged in the *tâtonnement*, thereby divorcing arguments concerning the 'convergence' to equilibrium from the 'variations' in prices, in terms of a change in the initial data.

This interpretation is consistent with the evidence contained in the *Elements*. It also implies that Walras is consistent in his construction of the argument. A sudden recognition of path-dependency in Part IV, where 'tickets' are first introduced, could only mean that Walras erred in not acknowledging that a similar device is also needed in Part III (the theory of exchange) to exclude re-distributional effects. The point is that 'tickets' were not designed for that purpose in the first place.

Notes to Chapter 2

1 Once equilibrium is obtained in Part V, however, the system is *stationary*: relative prices, quantities, and the rate of return remain constant over time, with net savings equal to zero. See Chapter 3 below.

2 Both Newman (1965:102) and Jaffé (1983:222–3) interpret this section of the *Elements* as allowing for the possibility of trades at false prices. In a later work, Jaffé (1983:247) provides a '*mea culpa*' for the error since, especially after the second edition of the *Elements*, such trades were explicitly excluded by Walras in Part II, as Jaffé (1954:566) indeed remarks in his collation notes to the fourth edition.

Patinkin (1965:522) and Walker (1970:691) note a 'change of mind' by Hicks on the issue. In Hicks (1934:342), it is claimed that Walras 'does not really make clear' whether or not out-of-equilibrium trades are allowed and remarks: 'If there is no actual exchange until the equilibrium prices are reached by bidding, then Walras's argument is beyond reproach on the score of logical consistency, though it may be called unrealistic.' In Hicks (1946:128), the claim changes to one in which Walras (as well as Edgeworth) 'supposed that demand and supply analysis ought strictly to be confined to such markets as permitted of "recontract", i.e., markets such that if a transaction was put through at "false" prices (we shall find it convenient to mark prices other than the equilibrium price), it could be revised when the equilibrium price was reached. Since such markets are highly exceptional, their solution of the problem (if it can be called one) was not very convincing.' As I will show, Hicks's interpretation is correct only with regard to Lesson 5, Part II of the *Elements*. In Part III, Walras does recognize that disequilibrium trades, without recontract, do occur (see Section 2.2 below), but he does not see the resulting possibility of path-dependent equilibria.

3 All the quotes in the above paragraph were first added in that edition.

4 In an earlier work, Walker (1970:692n15) in fact argues that the 'means by which Bertrand tried to demonstrate his contentions were ill chosen. He appears to have confused the mathematical indeterminacy that disequilibrium transactions introduce into a system of Walrasian static equations with a supposed indeterminacy of real markets. Bertrand brings an additional buyer into the market "at the last minute", which actually changes a market-day parameter. He seemed to believe that the market price "is indeterminate because it depends on accidental oscillations". And there are other peculiarities, not to mention nonsensical aspects, of this review.' The quotes which Walker uses come from Jaffé's (1965:790) summary of Bertrand's review.

5 Some of Walras's notation is changed where convenient.

6 Walras (1954:54) takes demand curves as downward-sloping throughout, i.e., as a matter of fact.

7 For the latter, in particular, the 'Law of Variation in Commodity Prices', there is an important point to note. Throughout the *Elements*, Walras rigorously separates the analysis of 'convergence' to a given equilibrium from the exercise of comparing two equilibrium points; a distinction that is especially relevant in Parts IV and V, where changes in aggregate data are excluded in the *tâtonnement*, so that any comparative-statics effects can be specifically ignored. See Sections 2.3 and 2.4 below.

8 Note Walras (1954:157): 'It should be recalled, moreover, that what we have in mind throughout this volume is not to pose and solve the problem in question as if it were a real problem in a given concrete situation, but solely to formulate scientifically the nature of the problem which actually arises in the market where it is solved empirically.'

9 Note Walker (1970:693): Walras 'never saw that it is not analytically defensible just to assume that disequilibrium transactions do not occur'; and Walker (1987a:760): 'Observing that disequilibrium transactions were not allowed in the nineteenth-century bourse, [Walras] incorporated that rule into his model of exchange.' Although Walker is correct with respect to Part II of the *Elements*, as well as the *tâtonnement* for multi-market exchange in Part III, he cannot be correct, in light of Walras's discussion of arbitrage, for the theory of exchange markets in the *Elements* as a whole.

Note Jaffé (1983:254n30) as well: 'Walras, from the 2d ed. of the *Elements* on, explicitly excluded trading at "false prices" from the operations of a truly competitive exchange market.'

10 One of the excess demand functions is redundant [Walras (1954:169–70)] to an aggregate budget constraint. The fact that excess demand functions depend only on relative prices already guarantees that trades occur only at equilibrium values. See Negishi (1962:646–8).

11 Note Walras (1954:162–3): 'Now there remains only to show—and this is the essential point—that the problem of exchange for which we have just given a theoretical solution is the selfsame problem that is solved empirically on the market by the mechanism of free competition.'

Walras (1954:171) recognizes that an equilibrium may not exist if the demand for a good falls to zero before offers take on a positive value.

12 This is what Arrow and Hahn (1971:305–6) refer to as 'stability theory' in terms of a Gauss-Seidel method of solving a set of simultaneous equations. Cf. Negishi (1962:656–7).

13 See Walras (1954:246) as well.

14 Lesson 13 (1954:173–81), on 'The Law of the Variation of Commodity Prices', analyzes the comparative-statics effects on prices of a change in the initial amounts of Q_j.

15 The same criticism applies (again) to Lesson 14 on the 'Theorem of Equivalent Redistributions'.

16 On Bertrand, see Section 2.1 above.

17 As Walker (1987b:160) documents, there *is* some hint of path-dependency in Edgeworth (1925,II:281): 'When we advance from the simplest type of market to the complexities introduced by division of labour, it is seen to be no longer a straightforward problem in algebra or geometry, given the natures of all the parties, to find the terms to which they come. Here, even if we imagine ourselves in the possession of numerical data for the motives acting on each individual, we can hardly conceive it possible to deduce *a priori* the position of equilibrium towards which a system so complicated tends.' But this seems to have had no effect on Walras. It is conceivable that the point may have been interpreted as an argument appropriate only to the presence of *multiple-equilibria* with bilateral trade, as in Edgeworth (1967 [1881]). Here, just as in Part II of the *Elements*, the final point of equilibrium (even if path-independent), is contingent on initial conditions and cannot be known beforehand. See Negishi (1962:660–3).

18 See Edgeworth (1967:46) as well.

19 Although Walras (1954:431–59) discusses the nature of monopoly and imperfect competition, he clearly sees competition as the most general case; the case formally analyzed in Parts II to VII of the *Elements*, e.g., Walras (1954:86): 'We shall study value in exchange as it arises under such competitive conditions. Economists, generally speaking, have fallen all too frequently into the error of studying value in exchange under unusual conditions. They are always taking about diamonds, Raphael's paintings, and concert recitals given by famous singers . . . But logic demands that we consider general before special cases, and not the other way around. What physicist would deliberately pick cloudy weather for astronomical observations instead of taking advantage of a cloudless night.'

20 Bortkiewicz (1890:85) shares a similar view. See Walker (1987:162).

21 Wicksell (1954 [1883]:81–2) does see the point in his discussion of 'direct exchange' versus the opportunity for arbitrage: 'If initially only direct exchange is permitted, but subsequently the market is entirely freed, each of the exchanging persons will acquire a greater profit by the wholesale trade or stock-exchange operations which will then take place, and in this way the total profit also can become greater. But the state of equilibrium thus attained will generally be different from that which would occur if trade were entirely free *from the beginning*.' Throughout the remainder of the *Lectures*, however, Wicksell (1934, I [1911]:64–5) simply assumes, 'under ideal market conditions', that the 'final price equilibrium is established from the very beginning'.

Wicksteed (1946 [1933]:226–7) is especially clear on the matter: 'Even if [traders] met at once (before they had provided themselves with substitutes or made any other consequential modifications in their other purchases or plans) and exchange among themselves till there was complete equilibrium, that final equilibrium would not exactly correspond with that which would have been established had the real conditions of the market been realized from the first. For the purchasers who bought at high prices, having forfeited a disproportionate amount of their resources, will be poorer, and those who had received a disproportionately large amount of the commodity will be richer, than they otherwise would have been . . . Now the mere distribution of wealth, the taking from one man's general resources and adding to another, essentially modifies the individual scales. Cobbets are not bidders for fancy pug-dogs or rubies, and Crœsus is probably not a bidder for fustian cloth or tripe. . . . Anything, therefore, which increases the total resources of some members of the community, and diminishes those of others, will *pro tanto* affect their estimates of the relative significance of different commodities. This will alter the elements of the communal scale of preferences, and the equilibrating price of any article will be affected, even though the tastes of the community and the total amount of the commodity remain the same.'

22 This is true even though Walras (1954:171), in Lesson 12, writes offer functions as multi-valued.

23 It may be argued that since Walras discusses stability in Part II of the *Elements*, there is no point in bringing the issue up again in the later sections. But note Walras (1936:352n): 'The theoretical equilibrium price is essentially a unique price, resulting at a given moment, in a general exchange.'

The usual interpretation in the secondary literature misses this. See, for example, Negishi (1962) and Walker (1987). Patinkin (1965) is the only exception, seeing stability and *tâtonnement* as separate issues for Walras.

24 See Hicks (1956:10–15).

25 Jaffé (1983:238) makes the error of suggesting that additively separable utility functions guarantee that all goods are gross substitutes (i.e., an increase in the price of one good results in an *increase* in excess demands for all other goods), as defined by Metzler (1945); and hence, that the Walrasian system—an example is given in Arrow and Hurwicz (1958)—is stable. With additive separability alone, however, it may be the case that all other prices *fall* with an increase in the excess demand for one good, implying total instability.

26 This is especially pertinent in the process of 'groping' in capital formation. See Section 2.4 below.

27 See Walras (1954:250–3, 287–8, 293, 326–7) as well.

28 In the Preface to the fourth edition, Walras (1954:37) remarks: 'In the theory of production, I no longer represented the preliminary groping towards equilibrium as it takes place effectively, but I assumed, instead, that it was done *by means of tickets* and then carried this fiction through the remainder of the book.'

29 Walker (1987a:765) is correct to point out that Jaffé should have translated *bons* as 'pledges', with the 'emphasis on a commitment', rather than 'tickets', since Walras did not use the French term *billets*. Jaffé (1954:528), however, essentially acknowledges this in his 'Translator's Notes' to Lesson 20: 'These "*tickets*" or "*bons*", as Walras calls them, are provisional contracts to buy or sell given quantities of goods or services at stated prices. What makes them provisional is the stipulation that the contracts are binding only if the stated prices turn out to be equilibrium prices for the economy as a whole; otherwise they become null and void. The translation of "*bons*" in this sense as "tickets" is suggested by the fact that tickets often read "*bon pour* . . ." in French, or "good for . . ." in English.' See Jaffé (1983:234–5) as well.

30 At the least, Walras (1954:247) sees this process of 'groping' as mimicking the way in which *actual* markets work: 'this is precisely the sort of groping which takes place in the products market under conditions of free competition, as entrepreneurs increase or decrease their output according as they make profits or losses.' See Walras (1954:41, 255) as well.

Walras (1954:253–4) asserts that as the output of a good expands, the prices of *all* inputs used to produce this good increase. He misses the point that if relative input-coefficients vary across goods, even though overall costs of production rise, the price of some inputs may rise, while others fall, as output increases.

31 Walras (1954:244–6, 249–53). See Chapter 3 below.

32 This specific quote—the 'problem of the path'—comes from Jaffé's (1954:529) 'Translator's Notes' to the fourth edition of the *Elements*. The same argument is repeated in all of Jaffé's later works.

33 Recall, in his earlier work, that Walker (1970:692n15) sees that 'the means by which Bertrand tried to demonstrate his contentions were ill chosen. . . . And there are other peculiarities, not to mention nonsensical aspects, of this review.' See Section 2.1 above.

It is important to note that Walker (1987a:768) recognizes that 'Walras did not, incidentally, owe an awareness of the problem caused by disequilibrium transactions or production to a study of Edgeworth's theory of recontracting.' This greatly qualifies Walker's (1970:693) earlier opinion where he suggests (in part) that Edgeworth's criticism of the 'groping' process shows, from the *Correspondence*, that 'more than once Walras recognized that disequilibrium transactions would render his model of exchange indeterminate.'

34 Walker (1983:766) misrepresents Patinkin here regarding Walras's asymmetrical treatment of 'groping' in production versus exchange. Patinkin (1954:534) does err (as Jaffé (1983:240–1) also notes) in suggesting that Walras 'does *not* make a provision for consumers' to make 'use of tickets' in the production model. But the essential asymmetry that Patinkin is concerned with is not this, but the fact that if Walras *is* aware of a path-dependency problem in Part IV, the 'tickets' device should rightly appear in Part III of the *Elements* as well.

35 Again, out-of-equilibrium trades also change 'the given conditions' in the exchange model. If Walras is aware of endowment effects, why are 'pledges' not included in Part III of the *Elements*?

36 See Walras (1954:143–9, 173–81, 255–63, 307–12). Note Walras's (1954:146) remark, just prior to discussing the comparative-statics of two-good exchange: 'The theorist has the right to assume the underlying price determinants are invariant over the period he has chosen to use in his formation of the law of equilibrium prices. But, once this formulation has been completed, it is his duty to remember that the forces that underlie prices are by their nature variable, and consequently he must formulate the law of the variation of equilibrium prices. This now remains to be done. Fortunately, the second formulation follows immediately from the first, for the very forces underlying the establishment of prices are the very forces that underlie the variation of prices, viz. the utilities and the quantities possessed of the commodities considered. These forces are, therefore, the primary causes and conditions of the variation of prices.'

37 This apparently explains the complete absence of a theory of the firm in the *Elements*. See Walras's uncomfortable, if not confused, handling of the issue in Part IV (1954:225) and Appendix I (1954:474n).

38 On the supply-side, Walras (1954:292) indeed assumes that entrepreneurs 'know their costs of production' as well as their 'selling prices', so that 'they know in advance whether they will make a profit or loss.'

39 In fact, this is what Bortkiewicz (1890:86) sees (although he simply restates Walras's position) as the essential feature of the *tâtonnement* process, i.e., as a static analysis '*in the sense* that [Walras] assumes the quantities of products possessed as being constant quantities, and the utility functions as invariant.' See Walker (1987a). Bortkiewicz also (apparently) misses the problem of path-dependency.

40 This seems to be the implication of the 'likely case' in the 'groping' process described in the second edition of the *Elements*.

41 Walras (1954:274). In the fourth edition Walras relates the amounts of individual savings to the utility from 'perpetual net income', with per-unit price 1/r. See Chapter 3 below. The issue is ignored here.

42 It is important to note that Walras does *not* take the values of prices and rentals, determined in Part IV, as fixed and given throughout the construction of the Part V system. Changes in the relative amounts of each K_i, in other words, coincide with different values for P_j, P_i, R_l, R_n and R_k until uniform returns are established. See Chapter 3 below.

43 'Second order' effects, for Walras (1954:288) are straightforward here. He fixes, for the time being, the equilibrium amounts of new capital goods produced, consistent with equation (2.4.8). An increase in savings implies a fall in the demand for consumer goods, with a fall in all service prices used to produce them. But a fall in service prices simply lowers both sides of (2.4.9), i.e., both sides move in the same direction, so that the increase in r predominates. The value of **y**, again, remains constant throughout.

Note Walras (1954:289): 'This particular groping which we have just described actually takes place in the stock exchange, which is the market for new capital goods.'

44 'Second order' effects, from a fall in an own-service price, although clearly affecting costs of production and selling prices of all other capital goods, again, for Walras (1954:293), 'cancel' each other out. A fall, in short, in an own-service price, cannot affect the values of all other rentals (since the relative outputs of some capital goods rise, while others fall) 'very much'; nor can it change r to 'any significant degree.'

45 See, for example, Lutz (1968) and Petri (1978). The usual response, in the secondary literature, is to re-interpret Walras as employing either an assumption of steady-state growth [e.g., Schumpeter (1954) and Montgomery (1971)], or static price-level expectations [e.g., Hicks (1934) and Diewert (1977)]. Neither of these, however, is consistent with the evidence in the *Elements*. See Chapter 3 below.

46 Likewise, for the 'feedback' from a change in r in equation (2.4.9), which changes S, to **y** and the demand for Q_j and K_i.

Chapter 3

The theory of stationary long-run equilibrium in Walras

This chapter deals with Walras's theory of long-run equilibrium. As I noted in Chapter 1 above, the secondary literature is especially critical of Walras in this regard. There are two distinct views. The first argues that Walras's procedure of using net savings to change the relative amounts of each capital good, until returns are equalized, is incompatible with his (purported) assertion that the prices and rentals of capital goods remain unchanged throughout. To avoid this error the *Elements*, by these accounts, must be *re*-interpreted as a system characterized by either steady-state growth, with constant prices, or by static price-level expectations, with actual (but not expected) returns varying over a sequence of (non-steady-state growth) periods. In spite of Walras's expressed procedure, in other words, either initial endowments cannot be arbitrarily given or expected and actual prices must vary.

The second claims that Walras, contrary to his stated intentions, simply errs in his attempt to derive a long-run equilibrium with uniform returns, while taking the initial endowment of each capital good as arbitrarily fixed and given. From this perspective, the best that Walrasian theory can obtain is a short-run equilibrium, Part V of the *Elements* notwithstanding.

For clarity, Sections 3.1 and 3.2 below set out—using a standard notation and approach—equations for long-run equilibrium as a formal analog to the Walrasian system. The complexities of Walras's actual equations (discussed later) are much easier to understand in comparison. Section 3.2, in particular, along with a further discussion of the secondary literature, shows that Walras's Part IV system is comparable to a long-run, one-sector, neoclassical theory of distribution, where each non-produced input receives an input-specific return.

Sections 3.3 and 3.4 examine, in detail, Parts IV and V of the *Elements*. Section 3.4a, notably, draws an important distinction—the absence of which adds to the confusion in the secondary literature—between a *preliminary* and Walras's *main* system of equations for long-run equilibrium with uniform returns. Section 3.4c sets out the bulk of the case against steady-state growth and static price-level expectations in the *Elements*, while 3.4d, in sharp contrast to standard accounts, considers the Part V system as a consistent and stationary long-run equilibrium. Sections 3.3b and 3.4b discuss the *tâtonnement* for the theory of production and capital accumulation and 3.5 concludes.

3.1 Equilibrium equations

Begin with the structure of equations and unknowns that define a standard long-run equilibrium. Since the system is static the resulting endogenous variables, given a set of parameters or initial data, do not functionally depend on time. If variables do have any sort of intertemporal dimension it is understood to be simply the result of a change in scale, not relative amounts.[1] With relative prices, quantities, income and consumption all constant over time, there is no need to date them; time subscripts are unnecessary.

To characterize this kind of system, in a simple way, assume the production of two (pure) consumption goods Q_a and Q_b, indexed by j, and represented by

$$Q_j = F_j(K_{1j}, K_{2j}, N_j, L_j) \tag{3.1.1}$$

for two produced capital goods K_1 and K_2, indexed by i, for convenience, N_j the amount of labour, and L_j the amount of land used to produce Q_j. For simplicity, think of K_i along Austrian lines as being produced by 'original' factors, labour and land, only[2]

$$\dot{K}_i + \delta K_i = G_i(N_i, L_i) \tag{3.1.2}$$

for $\dot{K}_i$ net additions and δK_i replacement output for each capital good.[3] Define Q_i as $(\dot{K}_i + \delta K_i)$, or gross investment. If long-run equilibrium is stationary, net additions to each K_i are identically equal to zero. Uniform returns on produced inputs [equation (3.1.10)], give an equilibrium structure of the capital stock such that the resulting production of each capital good is just sufficient to cover depreciation [$\dot{K}_i = 0$ or $\delta K_i = Q_i$ in (3.1.2)]. Both the absolute (physical) amount of each K_i, as well as the ratio of K_1 to K_2, remain intact. Once produced, the amount of each K_i

$$K_i = K_{ia} + K_{ib} \tag{3.1.3}$$

is assumed to be fully used in the production of Q_j.

To make the distinction between produced and 'original' factors especially clear, write

$$\begin{aligned} \bar{N} &= N_a + N_b + N_1 + N_2 \\ \bar{L} &= L_a + L_b + L_1 + L_2 \end{aligned} \tag{3.1.4}$$

so that the total amounts of (fully employed) labour and land are fixed data.[4] Taken separately, the stocks of labour and land are homogeneous; the stock of capital is not. Now derive first-order conditions for profit maximization in the usual way. Imagine a number of identical, single-product competitive firms producing $\mathbf{Q} = (Q_i, Q_j)$, taking all prices as parametric. Define $(\mathbf{P}, \mathbf{W}, \mathbf{R}) = (P_i, P_j, W, R, \mathit{R}_i)$ as the current-value or undiscounted price of output, the wage rate, the rental for land, and net rentals for each K_i respectively. Profit equations become

$$\begin{aligned} \mathbf{\Pi_j} &= P_jQ_j - \delta(P_1K_{1j} + P_2K_{2j}) - \mathit{R}_1K_{1j} - \mathit{R}_2K_{2j} - WN_j - RL_j \\ \mathbf{\Pi_i} &= P_iQ_i - WN_i - RL_i \end{aligned} \tag{3.1.5}$$

with the following first-order conditions

$$\begin{matrix} P_jF_{1j} - \delta P_1 = \mathit{R}_1 & P_jF_{Nj} = W & P_iG_{Ni} = W \\ P_jF_{2j} - \delta P_2 = \mathit{R}_2 & P_jF_{Lj} = R & P_iG_{Li} = R \end{matrix} \tag{3.1.6}$$

for $F_{1j} = \partial F_j/\partial K_1$, $G_{Ni} = \partial G_i/\partial N_i$ and so on. Let $R_1 = P_jF_{1j}$ and $R_2 = P_jF_{2j}$ be gross rentals, or simply the value of the marginal product of each K_i.

Constant returns to scale guarantees product-exhaustion and gives price equations, in per unit terms, of the form

$$\begin{aligned} P_j &= \mathit{R}_1k_{1j} + \mathit{R}_2k_{2j} + Wn_j + Rl_j \\ P_i &= Wn_i + Rl_i \end{aligned} \tag{3.1.7}$$

for $k_{ij} = K_{ij}/Q_j$, $n_i = N_i/Q_i$, and so on. Define input coefficients as $(\mathbf{k}, \mathbf{n}, \mathbf{l}) = (k_{1j}, k_{2j}, n_i, n_j, l_i, l_j)$. Equations (3.1.6) determine the optimal allocation of $(\mathbf{k}, \mathbf{n}, \mathbf{l})$ across $\mathbf{Q}$ and, with homogeneity of degree one, imply unit profit functions

$$\pi(\mathbf{P}, \mathbf{W}, \mathbf{R}) = 0 \tag{3.1.8}$$

for all firms, making (3.1.7) redundant to first-order conditions.[5]

Long-run equilibrium requires uniform returns over produced (and marketable) inputs, as assets. The condition ensures that savers are indifferent to holding, in this case, claims to ownership over K_1 and K_2. For the non-stationary case, portfolio equilibrium is given by

$$i = \frac{R_i}{P_i} - \frac{\delta P_i}{P_i} + \frac{\dot{P}_i}{P_i} \equiv r_i + \gamma_i \tag{3.1.9}$$

for i a 'nominal' rate of interest, r_i the rate of net income on each K_i, and $\gamma_i \equiv \dot{P}_i/P_i$, for $\dot{P} = dP/dt$, a capital gains/losses term.[6] If the system is stationary (or in steady-state growth), $\gamma_i = 0$ so that

$$i = r = r_i \tag{3.1.10}$$

for r a uniform real rate of return.[7] Note that both constant prices and unit depreciation (i.e., $\delta = 1$), gives: $(1 + r) = R_i/P_i$.

Finally, since there is more than one consumer good, write

$$ED_a(\mathbf{P}, \mathbf{W}, \mathbf{R}) = D_a - Q_a = 0 \tag{3.1.11}$$

as the excess demand for Q_a.[8] Equations (3.1.1)–(3.1.4), (3.1.6), (3.1.10) and (3.1.11) are twenty-three in number (i, j = 1, 2) and form an independent system. There are twenty-five unknowns: $(\mathbf{P}, \mathbf{W}, \mathbf{R})$, Q_j, K_i, K_{ij}, N_j, L_j, N_i, L_i, and the uniform rate of return r. The amount δK_i defines Q_i. If all prices are strictly positive, the choice of *numéraire* is arbitrary. Set $P_b \equiv 1$, expressing all prices relative to P_b. The system is indeterminate or short an equation. Only, say, for a given value of r, with $\bar{N}$ and $\bar{L}$ as parameters, can one, potentially, solve for the remaining unknowns.[9] With 'one degree of freedom', the exercise involves finding the 'missing equation', or set of equations and unknowns, to determine r.[10]

3.2 Walras

For Walras, a long-run system (with capital goods as produced inputs) is closed by adding an equation which expresses an equality between net savings and the total value of new additions to each K_i. The *initial* context—that is, until long-run stationary equilibrium with uniform returns is finally established—is one of non-steady-state growth. Walras, in fact, draws a distinction between a static, long-run system [Part IV of the *Elements*], where returns are input-specific, and a long-run equilibrium [Part V] where a uniform rate of return on reproducible assets is generated through net accumulation.

Begin with Part IV, the static system. By design, capital goods are neither produced nor traded, establishing only rentals, and not prices, for each K_i. Output consists of consumer goods only. The initial endowment of each type of capital good is arbitrarily fixed, listed now as 'original' factors of production, concurrent with land and labour. No input depreciates with use. Since the value of income is assumed to equal the value of consumption expenditures, individual savings (with no distinction between gross and net) are necessarily zero. There is, in short, a demand for the services of capital goods, but no demand for capital goods *per se*.

In terms of Section 3.1, equation (3.1.1) and the price equations for Q_j, in (3.1.7), remain unchanged. Economic profits are still zero. Since capital goods are not produced, first-order conditions appear simply as

$$P_j F_{ij} = R_i \quad P_j F_{Nj} = W \quad P_j F_{Lj} = R \tag{3.2.1}$$

with no rate of depreciation δ.[11] Equation (3.1.11) stays the same, except that excess demands no longer have P_i as arguments. Resource constraints, however, become

$$\bar{N} = N_a + N_b \qquad \bar{L} = L_a + L_b \qquad \bar{K}_i = K_{ia} + K_{ib} \tag{3.2.2}$$

so that each capital good is on the same footing as labour and land; as given endowments, rather than produced outputs. All other equations disappear. There are no prices—and no need for prices—for capital goods.[12] There is no uniform rate of return across different inputs. Equations (3.1.1), (3.1.11), (3.2.1) and (3.2.2) are fifteen in number (i, j = 1, 2) to determine sixteen unknowns: Q_j, K_{ij}, N_j, L_j, R_i, W, R, and P_j. Setting $P_b \equiv 1$, as before, closes the system.

Note that adding more consumer goods or inputs does not, for all essential purposes, change the nature of the system.[13] A Q_c, for example, adds six equations: a production function, four first-order conditions and an excess demand function; and six unknowns: Q_c, K_{ic}, N_c, L_c and a relative price P_c. A K_{3j}, or, as in the *Elements*, any added distinction between different types of labour or land, gives three new equations: a resource constraint

$$\bar{K}_3 = K_{3a} + K_{3b} \qquad (3.2.3)$$

and two first-order conditions; and three unknowns: K_{3a}, K_{3b}, and R_3.

Note, in particular, the effect of assuming a single produced consumption good and only one capital good (i, j = 1 above), as an 'original' input. Equation (3.1.1) simply becomes

$$Q = F(K, N, L) \qquad (3.2.4)$$

with resource constraints

$$\bar{K} = K \qquad \bar{N} = N \qquad \bar{L} = L \qquad (3.2.5)$$

and first-order conditions

$$F_K = R_K \qquad F_N = W \qquad F_L = R \qquad (3.2.6)$$

for R_K the rental for K. With only one consumer good there are no relative prices for output, but homogeneity of degree one in the function F guarantees

$$Q = F_K K + F_N N + F_L L = R_K \bar{K} + W\bar{N} + R\bar{L} \qquad (3.2.7)$$

or product exhaustion. Walras's Part IV equations, in other words, are reducible to a long-run, one-sector neoclassical theory of production and distribution.[14]

In Part V of the *Elements*, capital as well as consumer goods are produced, where the former (both old and new) now depreciate at rate δ with use. The stock of each $\bar{K}_i$, however, as endowments in Part IV, is taken as fixed and given in Part V. Newly produced capital goods, as net accumulations, enter the economy 'later', or in a 'period subsequent to the one under consideration'. Depreciation allowances, as new capital good production earmarked for replacement, enter 'now', so to speak, in Part V, to keep the amount of each $\bar{K}_i$ constant.[15] Assume, as well, a market where savings are allocated over ownership claims to capital goods as sources of perpetual net income in equilibrium. Changes in the quantities of each K_i produced, their rentals and asset prices, generate uniform rates of return.

Begin, as does Walras, by adding equations and unknowns to the Part IV system. Equation (3.1.2) defines the technology for the production of K_i ($i = 1, 2$), with the appropriate first-order conditions

$$P_iG_{Ni} = W \quad P_iG_{Li} = R \tag{3.2.8}$$

for the optimal allocation of N_i and L_i. Assume that newly-produced capital goods are identical, in physical form, to old capital goods, i.e., there is no technological change. Per unit price equations

$$P_i = Wn_i + Rl_i \tag{3.2.9}$$

give zero economic profits. In equilibrium, the 'rate of net income'

$$r = R_i/P_i - \delta \tag{3.2.10}$$

identical to equation (3.1.10), must be the same for all capital goods, old and new.

Let the value of gross savings

$$\mathbf{S} = \mathbf{S}(P_j, R_i, W, R, r) \tag{3.2.11}$$

be a function of consumer prices, all incomes and r. The value of new capital goods produced, or gross investment, is given by

$$I = P_1Q_1 + P_2Q_2 \tag{3.2.12}$$

for $Q_i = \dot{K}_i + \delta K_i$. Define

$$S = \mathbf{S} - \delta P_1K_1 - \delta P_2K_2 \tag{3.2.13}$$

as net savings. Walras, at first, restricts S to be strictly positive, with

$$\dot{K}_i = Q_i - \delta K_i \geqslant 0 \tag{3.2.14}$$

for each K_i, and $\dot{K}_i > 0$ for at least one capital good. Market equilibrium, finally, requires

$$S = P_1\dot{K}_1 + P_2\dot{K}_2 \tag{3.2.15}$$

or that net savings equals investment. Equations (3.1.2), (3.2.10), (3.2.11), (3.2.13) and (3.2.15) form a *preliminary* Part V system, adding, to Part IV, eleven equations and eleven unknowns: $\dot{K}_i$, N_i, L_i, P_i, $\mathbf{S}$, S and r.[16] The Part IV endowments of K_i determine the amounts $\delta\bar{K}_1$ and $\delta\bar{K}_2$ in Q_i. The remainder are net additions $\dot{K}_1$ and $\dot{K}_2$.

Note, in the above, that the unknowns P_j, W, R and R_i, were taken as given. The final step is to solve the Part IV equations, together with the preliminary Part V equations, simultaneously, i.e., to construct the *main* Part V system. The reasoning is clear. The transfer of resources from consumer to capital good production, since the functions F_j and G_i are non-identical, changes the value of prices and rentals. These change the value of savings and the mix of final output. Resource constraints must be redefined, and so on. Drawing on the equations from Section 3.1, the complete system becomes: (3.1.1), (3.1.2), (3.1.4), (3.1.6), (3.1.10), (3.1.11), (3.2.11), (3.2.13), (3.2.15) and

$$\bar{K}_i = K_{ia} + K_{ib} \qquad (3.2.16)$$

or twenty-six independent equations in twenty-seven unknowns: (**P**, **W**, **R**), Q_j, K_{ij}, N_j, $\dot{K}_i$, L_j, N_i, L_i, **S**, S and r. Setting $P_a \equiv 1$ closes the system. The asset-price for land, assumed 'fixed and indestructible', is given by

$$P_L = R/r \qquad (3.2.17)$$

while

$$K = P_1\bar{K}_1 + P_2\bar{K}_2 \qquad (3.2.18)$$

defines the value of the stock of endowed capital; adding two equations and two unknowns: P_L and K.

However, once long-run equilibrium is obtained for Walras—I set out the detailed case in Sections 3.4c and 3.4d below—the system is stationary: each $\dot{K}_i$ as well as net savings is zero. A uniform rate of return is generated through net accumulation, but once established, the equilibrium amounts of new capital goods produced must simply cover the depreciation of the fully adjusted stock of capital, leaving its structure intact.

The secondary literature has been largely concerned with two *apparent* errors in Walras's theory of long-run equilibrium: (a) Walras's practice of taking service prices, determined in Part IV, as fixed and given in Part V, i.e., as data in equations (3.1.2) (3.2.9), (3.2.10), (3.2.12) and (3.2.14), even though the numbers of each capital good are changing, or are about to change in a subsequent period; and (b) Walras's assertion that the prices of old capital goods, from the pre-existing list in Part IV, are the same as those of new capital goods produced in Part V.

Indeed, much of the commentary interprets (what I have termed) Walras's *preliminary* equations for long-run equilibrium as the *main*, or in fact the only, long-run system contained in Part V of the *Elements*. In this limited sense alone (i.e., given that such interpretations were correct), the problems are clear.

With regard to service pricing, for example, consider the requirement of uniform returns in stationary long-run equilibrium:

$$r = (R_i - \delta P_i)/P_i. \tag{3.2.19}$$

Fixing each R_i, as a result of a 'production equilibrium' in Part IV, with arbitrary amounts of $\bar{K}_i$, is, from this point of view, (rightly) inconsistent with the assumed context of positive net savings in Part V. As Lutz (1968:82) remarks:

A shortcoming of [Walras's] theory seems to me due to the fact that he transfers the tools of static equilibrium analysis to the analysis of a progressive economy without making the necessary modifications. Thus, he assumes that the prices of the services of new capital goods, which must be capitalized to obtain the prices of the capital goods themselves, remain forever at the level which establishes itself during the unit period in question. But in an economy which is growing as a result of net investments, relative prices must be changing from one period to the next, unless we make the impossible assumption (which Walras does not make) that all productive factors, including land, are increasing at a uniform rate. In an economy where only capital is increasing, the supposition that the prices of the services of capital goods will forever remain the same is not justified. Static assumptions are simply not applicable to a dynamic economy.

And, in the same manner, Petri (1978:254):

The root of Walras's difficulty can be traced to a contradictory attitude. Walras preserves the long-period condition of a uniform rate of interest over the supply prices of capital goods. This condition implies that prices are not changing through time (or changing slowly enough for the assumption of unchanging prices still to be a good approximation), as can be seen from the fact that the uniform rate of return on investment is determined by looking at the rentals of already existing capital goods—these rentals therefore are implicitly expected to be the same as those which new capital goods will earn. Now, this assumption of unchanging prices is unjustifiable in a model in which, like the Walrasian one, the endowments of capital goods are allowed to vary drastically from one period to the next.

The production of new capital goods, in other words, must, in general, change the values of R_i; either as a result—as Walras in fact argues—of changes in marginal products [Part VII of the *Elements*], or by the simple assertion [Part IV] of a negatively sloped demand function for the service price of each capital good in terms of its relative quantity.

Moreover, new capital good production must also generally change the values of P_j, W and R, as resources are transferred from consumer to capital good output. (Since Walras writes cost functions as increasing in service prices, fixed and identical input coefficients, across the production of various goods are evidently not assumed.) With P_j, W, R, and R_i all changing, the value of savings $\mathbf{S}(\cdot)$, in equation (3.2.11), in turn, cannot be defined with Part IV variables held constant.

A similar inconsistency applies to the prices of capital goods in the *preliminary* long-run system. Consider, for example, a case where $R_1 > R_2$ in Part IV. Positive net savings will presumably result in an increase in the production of K_1 relative to K_2. But since Walras clearly assumes that costs of production are increasing functions of output, the ratio P_1/P_2, in terms of new capital goods produced, must rise relative to the *numèraire*. Again, Lutz (1968:84):

> A fundamental difficulty remains which we have encountered several times already: one of the data in the Walrasian system consists of existing capital goods or, rather, of the supply curve for the services yielded by these. But if equilibrium is characterized by the condition that there be an equality between the prices of new capital goods and the prices of old capital goods plus depreciation for an appropriate number of years, then it must evidently be assumed that the stock of existing capital goods has from the beginning the very composition which would make this result possible. Just as Böhm-Bawerk and his successors had to assume that the time structure of the existing subsistence fund is exactly such as is necessitated by the equilibrium to be determined, so Walras has to assume the composition of the stock of durable capital goods to be exactly that which equilibrium requires. If Walras had really started, as undoubtedly he intended to do, with a completely 'random' composition of capital goods K′, K″, etc.—as he did and could do in the case of land and labour—then it would be most unlikely that all capital goods could and would in fact be replaced at his (temporary) equilibrium set of prices.

Arbitrary amounts of inital $\bar{K}_i$ and net accumlation, in other words, imply that the prices of old capital goods, however imputed, cannot be identical to the prices of new capital goods. These prices must vary in the 'move' from Parts IV to V of the *Elements*. Some holders of the pre-existing list of $\bar{K}_i$ will receive capital gains with accumulation; while others will experience capital losses.

As detailed in Chapter 1 above, various writers have interpreted Walras in a way that avoids (at least implicitly) these perceived problems in his theory of long-run equilibrium. The most common approach is to suggest that Walras has static price-level expectations in mind, e.g., Hicks (1934:346):[17]

'As Walras would have realized if it had not been for his confusion about the exact meaning of equilibrium, it is only in the stationary state that we can get any sensible sort of equilibrium, so long as people expect the prices of products to remain unchanged in the future (as Walras tacitly assumes they do).

And Diewert (1977:76):[18]

Producers and consumers expect that the prices which are established on the spot markets of the current period will also prevail in future periods. Thus Walras did not assume the existence of futures markets which would help consumers to intertemporally allocate resources but rather he assumed that economic agents formed *expectations* about spot prices in future periods based on current period spot prices, and these expected prices (along with the current interest rate) became a substitute for future prices.

Morishima (1980) is the most recent contributor to this view. Although the attribution to Walras of static price-level expectations is not explicitly made, Morishima (1980:553–4) indeed reads Walras as adopting a 'temporary equilibrium' approach; the approach typically identified with static price-level expectations.[19] The idea is to trace out a sequence of equilibria, where certain variables at t + 1, e.g., chosen additions to the stock of capital, presumably based upon expected prices, depend on decisions taken at time t. The growth process is seen as non-steady-state (with expected prices necessarily deviating from their actual values), just, so this view goes, as in Walras's 'move' from Parts IV to V of of the *Elements*.

Without steady-state growth, in other words, the only way (apparently) to interpret Walras's condition of a uniform rate of return, with positive net savings, is to assume that the expected value of γ_i in equation (3.1.9) is zero. The price of each capital good is simply expected to remain unchanged over time. The expected price of new capital goods, in particular, is assumed to be equal to the actual price of old ones. If the same is true for R_i, 'as Walras tacitly assumes', there is no problem with Walras taking these as given in Part V. The long-run system is consistently set out, as a 'temporary equilibrium'.

The alternative approach sees Walras as assuming steady-state growth. Equilibrium in Part V is not established, as Walras (1954:242) suggests, '*ab ovo*', by 'assuming the data of the problem to be any data whatsoever and yet invariable over a certain period of time', but instead with the 'right' amounts of each K_i in Part IV, or those already consistent with uniform long-run returns.[20] As Schumpeter (1954:1021n62) explains:

Stocks exist of course in specific forms, such as wine in cellars or handsaws in workshops. In reality, these specific goods need not at any given moment be, even approximately, what the maximizing conditions for the subsequent period would require. Here, so it seems to us now, we have the essentially dynamic problem of how the economic process adapts itself to situations that are inherited from the past and are always out of date when they are acted upon. But Walras eliminated this problem by the heroic assumption that stocks, like capital goods, are exactly as if they had been produced in the past with a view to conditions obtaining in the present. This is the meaning of the phrase *ab ovo*. We may render it by saying that he created an economic world in which every element fits perfectly into its niche, even if, owing to the fact that production takes time, it had to be produced when nobody could have known exactly what that niche would be like.

Montgomery (1971:293), more specifically, takes the 'move' from Parts IV to V of the *Elements* as:

the description of the one growth path on which the rate of change in prices always equals zero. From the deepest past this has been true, and true it will remain. Then all we need to assume about expectations is that everyone believes that prices will continue to be what they are.

And, again, Montgomery (1971:293):[21]

If this interpretation is accepted, the claim that Walras's theory of capitalization and credit is a growth model can be established. If we assume that a unique equilibrium solution to Walras's 'static' model exists, it will be sufficient to show that an equilibrium growth path exists and that on this path $\dot{P}_i = \dot{r} = \dot{R}_i = 0$. Although no argument about stability is possible, the definition of equilibrium assures that once an economy is on this path, it will remain there.

Not only, in other words, must the pre-existing amounts of $\bar{K}_i$ be 'right', but each capital good in Part V, as well as land and labour, must grow at the same proportional rate $g = \dot{K}_i/K_i$.

If all inputs grow at rate g, with net accumulation, all relative prices, in particular, P_i and R_i, will remain unchanged over time. The value of γ_i in equation (3.1.9), relative to the chosen *numèraire*, will be zero. All that is required is that each $\bar{K}_i$ in Part IV already exist in the relative amounts consistent with r—the meaning of '*ab ovo*' for Schumpeter. Walras's long-run system in Part V is consistently set out as a steady-state equilibrium.

3.3 **Part IV of the *Elements***

3.3a *Production equations*

With Sections 3.1 and 3.2 in mind, particularly the criticisms of Walras's theory of long-run equilibrium, we now turn (here, and in Section 3.4 below) to a detailed examination of the *Elements*.

Jaffé (1942:37) takes the position that Part IV of the *Elements* 'corresponds approximately to Marshall's "Equilibrium of Normal Supply and Demand" in the short period'.[22] But this cannot be the case. Although the amount of each capital good is arbitrarily fixed, it is fixed only for the system as a whole. Walras treats capital goods proper like any other natural resource, as non-augmentable, everlasting inputs whose services are freely transferable (albeit at increasing costs) between uses. The context is one of long-run cost price. Equilibrium is defined as a point where 'the selling prices of products equal the costs of the productive services that enter into them', so that 'entrepreneurs make neither profit nor loss'.[23] Although the *tâtonnement* implies both output and price adjustment in response to economic profits or losses, as a 'move' to a long-run position, there is nothing in the argument—in the production equations of Lessons 20–21 of the *Elements*—that resembles a Marshallian short-run equilibrium.

Walras (1954:237) begins in Lesson 20 with a heterogeneous list of productive services as 'the essential data of our problem', measured as 'a certain quantity of land-service per day from a hectare of such and such a piece of land; a certain quantity of labour per day from such and such a person; and a certain quantity of capital-service per day from such and such a capital good'. Only consumer goods are produced, with the application of inputs, the realization of output, and consumption all taking place simultaneously, or 'during the same period of time'. Capital goods do not depreciate [Walras (1954:238)] and there is no 'saving for the purpose of creating new capital goods'. The value of output, in short, equals the value of income, and all income is spent on consumer goods.

Define (**l**, **n**, **k**) = ($L_1, L_2 \ldots, N_1, N_2 \ldots, K_1, K_2 \ldots$) as the given amounts of each type of land, labour and capital good, or n different services in total, indexed by i where appropriate.[24] Walras designates Q_a, Q_b, Q_c, . . . as the output (or, in the proper context, the amount purchased or demanded) or m consumer goods, indexed by j, with prices P_j and rentals (R_l, R_n, R_k) for each service.

Individuals are assumed to possess utility functions defined over all services and goods, with the total offer of each service

$$O_i = F_i(P_j, R_1, R_n, R_k) \qquad (3.3.1)$$

and the total demand for each good

$$D_j = F_j(P_j, R_l, R_n, R_k) \qquad (3.3.2)$$

as functions of all prices. Individual budget constraints are given by

$$o_lR_l + o_nR_n + o_kR_k = d_a + d_bP_b + d_cP_c + \ldots \qquad (3.3.3)$$

for $P_a \equiv 1$, as *numéraire*.

Input coefficients (a_{lj}, a_{nj}, a_{kj}) for the per-unit production of each consumer good are fixed and given. For each (**l**, **n**, **k**), long-run equilibrium implies n equations of the form

$$O_k = a_{ka}D_a + a_{kb}D_b + a_{kc}D_c + \ldots \qquad (3.3.4)$$

so that '*the quantities of productive services used are equal to the quantities effectively offered*'; and

$$P_j = a_{lj}R_l + a_{nj}R_n + a_{kj}R_k \qquad (3.3.5)$$

or m equations 'expressing the fact that *the selling prices of the products are equal to the cost of the productive services* employed in their manufacture'.

The system (3.3.1)–(3.3.2) and (3.3.4)–(3.3.5) is closed, representing, for Walras (1954:241), 2m + 2n − 1 equations in the same number of unknowns: (a) n total quantities of services offered, i.e., as an expression of the demand for inputs to produce output; (b) the n prices or rentals of these services; (c) the m total quantities of each consumer good demanded; and (d) the m − 1 relative prices of these goods in terms of the *numéraire*.[25]

Most importantly, just as in the system of equations (3.1.1), (3.1.11), (3.2.1) and (3.2.2), there is no uniform rate of return on capital. Each capital good in Part IV of the *Elements* is treated on the same footing as the 'original' factors of production, land and labour. Each receives a specific rental R_k. The system is nevertheless in long-run equilibrium, in so far as equation (3.3.5) imply zero economic profits. But without net accumulation, for Walras, there is no rationale for the determination of the prices of individual capital goods or for a uniform rate of return.

3.3b Tâtonnement *in the theory of production*

In Part III of the *Elements* the quantity of each traded consumer good is taken as given—the 'essential data' of the problem—to solve for relative prices as an exchange equilibrium. Changes in the 'data', in turn, define a comparative-statics exercise and a new set of prices. In Part IV, however, it is the quantity of each service that is treated as exogenous; the produced amounts of Q_j are unknowns that depend on how services are allocated between uses.

The system 'converges' to the equilibrium defined in Section 3.3a above, by entrepreneurs responding to economic profits or losses. As Walras (1954:224–5) explains:

> Equilibrium is the normal state, in the sense that it is the state towards which things spontaneously tend under a regime of free competition in exchange and in production. In fact, under free competition, if the selling price of a product exceeds the cost of the productive services for certain firms and a *profit* results, entrepreneurs will flow towards this branch of production or expand their output, so that the quantity of the product (on the market) will increase, its price will fall, and the difference between price and cost will be reduced; and, if (on the contrary), the cost of the productive services exceeds the selling price for certain firms, so that a *loss* results, entrepreneurs will leave this branch of production or curtail their output, so that the quantity of the product (on the market) will decrease, its price will rise and the difference between price and cost will be reduced.

'The desire,' in short, 'to avoid losses and to make profits is the mainspring of the entrepreneur's actions in demanding productive services and offering products for sale.' Equilibrium occurs when the supply and demand for both goods and inputs are equal, with resulting prices for each good simply covering costs of production.[26]

It is at this point in the *Elements* that Walras (1954:242) introduces the well-known device of 'tickets' for the *tâtonnement* in the theory of production.[27] Equilibrium is established '*ab ovo*' (i.e., by 'assuming the data of the problem to be any data whatsoever and yet invariable over a certain period of time'), through the issue of 'provisional contracts', or 'pledges', for the quantities of goods and services demanded and supplied, until price adjustments guarantee a long-run outcome. Production and exchange, in other words, take place only when equilibrium prices are obtained.

The idea is to construct a formal analog to the *tâtonnement* contained in the pure theory of exchange, Part III of the *Elements*; a concern, in this case, with 'variations' around a given long-run equilibrium. Allowing, more precisely, for actual production at 'false' long-run prices would imply, for Walras, tracing out a sequence of short-run equilibria (as the production of each consumer good changes with economic profits or losses) as an exercise in comparative-statics. The 'tickets' device avoids this by finding the right mix of consumer good output from the outset—the analog to Part III—without actual production taking place. Walras, in effect, purposely avoids the Marshallian short-run by characterizing, instead, out-of-equilibrium movements around the system defined by equations (3.3.1)–(3.3.2) and (3.3.4)–(3.3.5).

In Lesson 21, Walras (1954:243) indeed begins the analysis *in* long-run equilibrium, with zero economic profits, and then imagines the effect of 'variations' in outputs on all prices and quantities. A distinction is drawn between the 'selling price' of output or P_j, and its cost-price or P_j^*, on the one hand, and 'ticket' amounts of each consumer good (to be produced) or Q_j, and its equilibrium value or Q_j^*, on the other.

In terms of standard supply and demand functions for a single market, for example, a $Q_b < Q_b^*$ implies (a) that P_b is greater than P_b^*, designating economic profits; (b) that the demand for services at Q_b is greater than their current 'offer'; and (c) that the current rental for each service, say R_l, is less than its equilibrium rate or R_l^*. As the output of Q_b expands for Walras (1954:243–54), its price falls (as a move along a given demand curve), costs of production rise (as a move along a product supply curve), and rentals for each service increase until offers and demands are equal, with zero profits.

The multi-market effects are quite complex and although Walras's analysis is imperfect, it will prove useful to review some of the argument. Walras starts on the product-side, writing

$$Q_j = F_j(P_a, P_b, P_c, \ldots, R_l, R_n, R_k) \qquad (3.3.6)$$

as the output of each good, holding service prices constant, where $Q_j \cdot (P_j - P_j^*) \gtrless 0$ defines economic profits or losses. In particular, the functions F_j are assumed by Walras (1954:244–5) to be increasing in P_j, so that if the selling price of Q_b, for example, exceeds its cost-price, the output of good b increases. An increase in Q_b, in turn, given all other prices for output, lowers P_b until $Q_b = Q_b^*$.

The question, however, is what happens to the market for Q_b if a fall in its price changes the demands for other goods—what Walras terms a 'second order' effect. Individual income

$$y = o_l R_l^* + o_n R_n^* + o_k R_k^* \qquad (3.3.7)$$

in terms of equilibrium service prices, is assumed to remain constant throughout. (Again, a formal analogy is made to Part III of the *Elements*, where the value of 'initial endowments' remains unchanged in the *tâtonnement* to exchange equilibrium.) The value of expenditures on consumer goods

$$y = d_a + d_b P_b + d_c P_c + \ldots \qquad (3.3.8)$$

totally exhausts income. Summing over individuals gives the total demand for each good.

The 'prime' effect of a fall in P_b is to increase the demand for Q_b. But clearly a change in 'each individual's outlay' on this good, or $d_b P_b$, changes the pattern of demands generally, with a change in prices altering the demand for Q_b again. To handle this, Walras (1954:246) simply assumes: (a) either that the spill-over in expenditure to other goods, from a change in the amount spent on Q_b, is relatively negligible; or (b) if such 'second order' spill-overs occur, to a large extent, they generate self-compensating effects on the total demand for Q_b. In particular, since some prices rise, presumably as complements to Q_b, while others fall, as substitutes, the total 'effect of these purchases and sales cancel each other out'.[28] The demand function for Q_b, in short, remains unchanged as its price falls. Performing the same operation on all goods, e.g., raising prices for products in which $P_j < P_j^*$, clears all markets at cost-prices.[29]

The same reasoning is used on the input-side. A $P_j > P_j^*$ implies, for Walras (1954:250), that 'ticket' demands for, say, land are less than the offers of this input

$$O_l = F_l(P_j, R_j, R_n, R_k) \qquad (3.3.9)$$

at given selling prices for goods and all other services. Costs are clearly [Walras (1954:249)] written as increasing in each service price, so that as the output of some particular Q_j increases, so does R_l until offers and demands are equal at R_l^*. But a change in R_l results in changes in offers, demands, and service prices for each labourer and capital good, with a feedback on the market for land. Walras (1954:251) simply assumes, once again, that such 'indirect influences', in terms of the effect on R_l^*, are 'cancelled' out. Since the outputs of some goods are rising, while others fall, the overall changes in demands for inputs are seen as 'more or less' off-setting. Adjustments continue to occur until (R_l^*, R_n^*, R_k^*) is obtained, with market clearing at $(\mathbf{l}, \mathbf{n}, \mathbf{k})$.

Although Walras (1954:253–4) sees costs as increasing in service prices, he asserts that an increase (decrease) in say Q_b, increases (decreases) the prices of all inputs used to produce this good. There seems to be little, if any, recognition that if relative input coefficients vary across goods, even though overall costs of production change, the price of some inputs may rise, while others fall, with a change in output. Moreover, the feedback from changes in output to changes in service prices and hence income, and from a change in income to a change in the demand for consumer goods, is ignored altogether. The measure of $\mathbf{y}$ remains constant in the *tâtonnement* to long-run equilibrium.

3.4 Part V of the *Elements*

3.4a *Equations for capital formation*

In Part V of the *Elements*, Walras extends the idea of long-run equilibrium to incorporate a uniform rate of return on capital goods as reproducible inputs, now depreciating with use. The stock of each $\bar{K}_i$ from Part IV, however, remains unchanged, with depreciation allowances simply re-invested in each to keep relative amounts intact. Positive net savings imply the production of *new* capital goods (say) K_i, effectively changing the relative amounts of each K_i, in order to generate both prices for each capital good (or P_i) and a uniform r. A key proviso is that the production of new capital goods, by design of the *tâtonnement* for capital formation, literally enter the system only 'later', or in a 'period subsequent to the one under consideration.'

Walras (1954:269) is clear that the production equations in Part IV *per se*, cannot solve for any P_i. With all income spent on consumption, no depreciation, and no rate of time preference, only rentals for each capital good are determined; only rentals matter [see Section 3.2 above]. What is needed is a market for the purchase and sale of each K_i, with the desire for some individuals wishing to forego present consumption, while others want to consume more than their current incomes, providing the rationale for such exchanges:

> If there is to be a demand for, a supply of, and prices of capital goods, we must suppose that there are land-owners, workers and capitalists who purchase consumers' goods and services in amounts that either fall short of or exceed their incomes, thus leaving them either a surplus with which to buy capital goods or a deficit which compels them to sell their capital goods. According as the excess of income over consumption is greater or less than the excess of consumption over income in the aggregate, an economy is either progressive or retrogressive.

Walras, as mentioned, restricts the analysis—in the 'move' from Parts IV to V, and until a long-run stationary equilibrium is established—to the 'progressive' case, i.e., to positive net savings in the aggregate, so that both consumer and capital goods are now produced in Part V. The context, however, remains one of long-run cost-price. The output of each K_i is varied until economic profits are zero.

Walras (1954:268–9) begins in Lesson 23, where produced and 'original' inputs are distinct, with the defining characteristic of a long-run equilibrium, i.e., a uniform rate of return

$$r = (R_i - \delta_i P_i)/P_i = \mathit{R}_i/P_i \qquad (3.4.1)$$

on each capital good, for δ_i the rate of depreciation and R_i a gross rental or income. 'Under certain normal and ideal conditions,' in other words, 'when the market for capital goods is in equilibrium', the prices of capital goods are 'rigorously proportional to their net incomes.'[30] Specific '*depreciation charges*', once again, are fully re-invested in each K_i, leaving each intact as a 'permanent' source of income.

As a first step, Walras simply adds equations and unknowns for capital formation to those already existing in the theory of production. In particular, the values Q_j, P_j, R_l, R_n and R_k from the Part IV system are provisionally taken as given in Part V to construct, as discussed in Section 3.2 above, a *preliminary* long-run equilibrium.

New capital goods, for Walras (1954:271), are produced at cost-prices adding i equations of the form

$$P_i = a_{li}R_l + a_{ni}R_n + a_{ki}R_k \tag{3.4.2}$$

for given amounts of (**l**, **n**, **k**). The response to economic profits and losses drives the result. Individual savings

$$\mathbf{s} = o_l R_l + o_n R_n + o_k R_k - [d_a + d_b P_b + d_c P_c + \ldots] \tag{3.4.3}$$

is defined as the difference between income and expenditures, where

$$\mathbf{s} > \delta_1 P_1 k_1 + \delta_2 P_2 k_2 + \delta_3 P_3 k_3 + \ldots \tag{3.4.4}$$

for k_i the individual holdings of each K_i, implies net accumulation, or an amount 'greater than the sum needed to cover the depreciation' of the 'existing capital goods proper' from Part IV.[31]

In the fourth edition of the *Elements*, Walras brings savings behaviour under the umbrella of individual optimization, by imagining the existence of 'a commodity (E) consisting of *perpetual net income*'.[32] The per-unit price of E is given by

$$P_E = 1/r \tag{3.4.5}$$

with $d_e P_E$ as the value of individual demands for net additions to perpetual net income in terms of purchases of new capital goods. The measure, in turn, of

$$q_e = \delta_i P_i k_i \tag{3.4.6}$$

summed over i, defines the amount that just covers depreciation, with

$$\phi_e(q_e + d_e)/P_E = \phi_a(d_a)/P_a \qquad (3.4.7)$$

as the 'condition of maximum satisfaction', in terms of the *numéraire*, for ϕ_e as the marginal utility from E.[33] Equation (3.4.7), for Walras, implies

$$d_e = F_e(P_j, R_l, R_n, R_k, P_1, P_2, \ldots, P_E) \qquad (3.4.8)$$

or a demand function for net additions to capital and, summing over individuals,

$$S = D_E P_E = F_e(P_j, R_1, R_n, R_k, P_i, P_E) \qquad (3.4.9)$$

as a measure, initially restricted to be positive, of 'aggregate excesses of income over consumption'. Capital market equilibrium, finally, requires

$$S = P_i K_1 + P_2 K_2 + P_3 K_3 + \ldots \qquad (3.4.10)$$

or an equality between net savings and the value of net investment.

The *preliminary* system is complete.[34] Equations (3.4.1), (3.4.2), (3.4.9) and (3.4.10) are 2i + 2 in number to solve for 2i + 2 unknowns: K_i or the 'quantities produced of new capital goods'; P_i or their relative prices in terms of the *numéraire*; S or the amount of net savings; and r or the uniform 'rate of net income'. Walras (1954:276), in turn, writes the system in reduced form by combining equations (3.4.1) and (3.4.2), solving for P_i, and (3.4.9) and (3.4.10), solving for S; giving

$$R_i/(r + \delta_i) = a_{1i}R_1 + a_{ni}R_n + a_{ki}R_k \qquad (3.4.11)$$

and

$$F_E(P_j, R_l, R_n, R_k, P_i, P_E) = P_1 K_1 + P_2 K_2 + P_3 K_3 + \ldots \qquad (3.4.12)$$

or i + 1 equations in i + 1 unknowns, K_i and r. If the ratio of net income, in particular, is not the same for any two capital goods, 'it will be advantageous to produce less of the capital good for which the ratio is smaller and more of the capital good for which this ratio is larger.' Equilibrium requires uniform returns on all produced inputs.

In Lesson 24 Walras combines the *preliminary* Part V system with the equations from Part IV, solving both simultaneously, to obtain the *main* Part V equations for long-run equilibrium. This is a crucial step.[35] As Walras (1954:276) explains:

> We might have proceeded with the aid of the above equations [(3.4.11) and (3.4.12)] to determine the i + 1 unknowns from which we could then have deduced the prices of the new capital goods and the aggregate value of savings, provided, however, that we neglected the resulting changes in the quantities of products to be manufactured and in the prices of products and services attributable to saving and capital formation. But our aim is to gain a comprehension of the economic mechanism as a whole. Consequently, in spite of the complexity of the notations (which is, after all, a minor, though unavoidable, inconvenience), we propose to combine into one system both the 2m + 2n − 1 production equations and the 2i + 2 equations of capital formation and credit.

The amounts K_i, in other words, must change the values—in a way made specific in the *tâtonnement*—of Q_j, P_j, R_l, R_n, and R_k as resources are transferred from the production of consumer to capital goods. These changes, in turn, affect the level of savings and the relative amounts of each K_i produced. The *main* equilibrium system is given by (3.3.1), (3.3.2), (3.3.4), (3.3.5), (3.4.1), (3.4.2), (3.4.9) and (3.4.10); or 2m + 2n + 2i + 1 equations in the same number of unknowns, with a '*uniformity of the rate of net income* for all capital goods proper.'

3.4b Tâtonnement *in the theory of capital formation*

Once again, the 'tickets' device is used to analyze 'variations' around a given long-run position, obtained '*ab ovo*' or with any given data. As mentioned, new capital goods, for Walras (1954:282–83), only enter the system 'later':

> Thus equilibrium in capital formation will first be established *in principle*. Then it will be established *effectively* by the reciprocal exchange between savings to be accumulated and new capitals to be supplied *within a given period of time*, during which *no change in the data is allowed*. Although the economy is becoming *progressive*, it remains (for the time being) *static* because of the fact that the new capital goods play no part in the economy until later in a period subsequent to the one under consideration.

The arbitrary amounts, in short, of (**l**, **n**, **k**), from Part IV, used now to produce both Q_j and K_i, remain constant in the *tâtonnement* to determine r.

Walras (1954:284–89) begins in Lesson 25 with 'random' service prices and quantities. In general, the resulting amount of S generates both

$$S \neq P_1K_1 + P_2K_2 + P_3K_3 + \ldots \qquad (3.4.13)$$

or an inequality between savings and investment and

$$P_i \neq R_i/(r + \delta_i) \qquad (3.4.14)$$

or prices for each capital good which are not consistent with uniform returns. The value of S, from equation (3.4.9) is assumed to be both increasing in r and each service price.[36]

Walras proceeds with equation (3.4.13) first, assuming an equality in (3.4.14), so that

$$S < [R_1/(r + \delta_1)]K_1 + [R_2/(r + \delta_2)]K_2 + \ldots \qquad (3.4.15)$$

or the value of savings, for example, is less than the value of net additions to the stock of capital. An equality in (3.4.15) is simply established through an increase in the rate of return, the 'prime' effect, decreasing the right-hand side and increasing the left.[37]

But clearly, once again, there are 'second order' effects. An increase in savings implies a fall in the demand for consumption goods and, for Walras, a fall in *all* service prices for inputs used to produce Q_j.[38] A fall in R_l, R_n and R_k, in turn, since each K_i is produced at cost-price [equation (3.4.2)], will change the value of savings and new capital good production. Walras's 'solution' is to fix, for the time being, the quantities of K_i that are consistent with equation (3.4.14). A fall in service prices, as a result, simply lowers both sides of (3.4.15) and since, for Walras (1954:288), both sides move in the same direction, the 'prime' effect of a fall in r predominates. With enough iterations, equation (3.4.15) comes into an equality with changes in both (R_l, R_n, R_k) and the rate of return.[39]

Next, Walras (1954:290–4) considers equation (3.4.14), allowing the output of each K_i to vary, assuming an equality in (3.4.13) between savings and investment. As in the *tâtonnement* for Part IV, the selling price for each capital good

$$P_i = R_i/(r + \delta_i) \qquad (3.4.16)$$

is distinguished from its cost-price or P_i^*, where $K_i \cdot (P_i - P_i^*)$ defines profits or losses for capital good producers. Take the case, for example, where

$$R_i/(r + \delta_i) > a_{li}R_l + a_{ni}R_n + a_{ki}R_k \qquad (3.4.17)$$

or $P_i > P_i^*$. Costs of production vary inversely with the output of each K_i. Own-service prices, e.g., R_1 on K_1, and hence selling prices are assumed to fall with an increase in the output of each capital good.[40] The presence of economic profits in equation (3.4.17), for Walras (1954:291–2), results in the 'prime' effect of a rise in (R_l, R_n, R_k) and a fall in R_i as K_i increases, resulting in the equality of P_i and P_i^*. For capital goods, on the other hand, in which $P_i < P_i^*$, i.e., effectively those with low rates of return, outputs and costs of production fall, while own-service prices rise. The relative amounts of each K_i change until r is earned on all capital goods.[41]

An increase in a particular K_i, however, generates a 'second order' effect, due to a fall in R_i and an increase in all other service prices, on the costs of production and prices of all other capital goods. Walras (1954:293) simply assumes (again) that such effects 'cancel' each other out. Since the relative outputs of some capital goods increase, while others decrease, the overall effect of a fall in R_i cannot change the values of (R_l, R_n, R_k) 'very much'; cannot change the rate of net income, and hence selling prices to 'any significant degree'.[42] The feedback from a change in service prices in equation (3.4.17), on the amount of savings defined in (3.4.15), is ignored altogether.

3.4c *The theory of long-run equilibrium in Part V*

It is quite clear, contrary to Lutz (1968) and Petri (1978), that Walras does *not* hold the service prices determined in Part IV of the *Elements* fixed throughout the construction of long-run equilibrium in Part V.[43] Although this indeed is done in the *preliminary* long-run system contained in Lesson 23, Walras does not end the discussion here. The whole point of Lesson 24 'is to gain a comprehension of the economic mechanism as a whole', to solve the equations from Parts IV and V simultaneously. The values of (R_l, R_n, R_k), in particular, are again listed among the 2m + 2n + 2i +1 unknowns, and 'all the unknowns', for Walras (1954:307), 'depend on all the equations of economic equilibrium'. Savings and capital accumulation *result in changes in Part IV variables*, 'in the quantities of products to be manufactured and in the prices of products and services'.[44]

This is especially evident in the *tâtonnement* for capital formation. As resources are transferred from consumer to capital good production, P_j, Q_j, R_l, R_n and R_k must adjust, as 'first order' or 'prime' effects, with the determination of K_i and P_i, until returns are equalized across all capital goods. Indeed, in Lesson 25 Walras (1954:291) explicitly writes

$$R_k = F_k(K_1, K_2, K_3, \ldots, P_E) \tag{3.4.18}$$

or the 'prices of services as functions of new capital goods to be manufactured and the price of net income'. (In equilibrium, net rentals R_k for each capital good are equal to $[R_i - \delta_i]$ in terms of new capital good production.)[45] The production of K_i, in short, alters the service prices of old capital goods, from the pre-existing list in Part IV. Lessons 22 and 25 make it clear that an increase in K_i results in a fall in R_i and an increase in the value of all other service prices.[46] If, for example, the return on K_1 is greater than K_2

$$(R_1 - \delta_1 P_1)/P_1 > (R_2 - \delta_2 P_2)/P_2 \tag{3.4.19}$$

the ratio of produced amounts of K_1/K_2 must increase, increasing P_1/P_2, as functions of (R_l, R_n, R_k), and decreasing R_1/R_2, until returns are equalized. Although, in a modern technical sense, the *tâtonnement* is obviously imperfect, the *tâtonnement*, on Walras's own account, requires these movements in prices and rentals to establish a uniform 'rate of net income'.

The same conclusion applies to P_i. Although Walras (1954:276) asserts that the prices of new capital goods 'are necessarily the same as the prices of existing capital goods proper', the *tâtonnement* for capital formation requires, again from Walras's own account, that the relative prices of capital goods change in the 'move' from Parts IV to V of the *Elements*. In other words, since costs of production are seen as increasing in K_i, a change in the relative amounts of each K_i must change relative prices.

This in fact is the whole point of Walras's distinction between P_i (selling prices) and P_i (cost-prices). A $P_i > \mathrm{P}_i$ implies both economic profits and a relatively high rate of return on K_i, or

$$P_i = R_i/(r + \delta_i) > \mathrm{P}_i = a_{li}R_l + a_{ni}R_n + a_{ki}R_k \tag{3.4.20}$$

in terms of equations (3.4.16) and (3.4.17). The resulting increase in K_i decreases its selling price and raises overall costs of production. This could not be more clear. The only way for P_i to fall—recall, at this point, that Walras takes r as given from an equality between savings and investment in equation (3.4.13)—is for R_i to fall with an increase in K_i. The only way for P_i to rise is by an overall increase in (R_l, R_n, R_k). A change in the relative price of each capital good is part of the *tâtonnement* which obtains r.

Read in this context, Walras's claim that the prices of old and new capital goods are identical is easy to understand. Walras (1954:309), in Lesson 28, uses the device of imputing prices to the pre-existing list of K_i, in Part IV, only after the equations for long-run equilibrium in Part V have been solved; only after, in other words, the process of adjusting rentals, prices, and the quantities of new capital goods in the *tâtonnement* for capital formation, along with the amount of savings and r, has been completed:[47]

> The rate of net income and the prices of new capital goods having once been determined, the prices of all existing capital goods of whatever category, be they capital goods proper, landed capital or personal capital, are determined *ipso facto*, for the net incomes from these goods are simply quantities of the commodity (E) the price of which is determined, as it happens, by the equality between its offer and the demand for it based on the pursuit of maximum satisfaction [equation (3.4.7)] in the light of the initial quantities possessed.

And, from the Preface to the fourth edition of the *Elements*, assuming, for convenience, no depreciation:

> Once we have the rate of income (from capital goods), we can obtain not only the price of newly manufactured fixed capital goods, but also the price of old fixed capital goods (viz. land, personal facilities, and already existing capital goods proper) by dividing the prices of their services (i.e., rent, wages, and interest) by the rate of income. (1954:42)

Since, by assumption, old and new capital goods are physically identical, their prices (ignoring depreciation) must be the same, once long-run equilibrium in Part V is obtained. The values of R_k, and hence the prices for old capital goods, cannot be known until the main long-run system has been solved; in particular, they cannot be known until the values of P_i, R_i (equal to $[R_k + \delta_k]$ in equilibrium), S and r are also determined.[48]

All of this should also make it clear, contrary to Schumpeter (1954), Montgomery (1971) and Kuenne (1963), that Walras does not adopt a steady-state assumption, i.e., that he does not construct a list of pre-existing capital goods in Part IV, in the exact proportions already consistent with long-run equilibrium in Part V. (If nothing else, Walras's (1954:386–91) recognition of land as the ultimate fixed factor of production makes this impossible.) The term '*ab ovo*', throughout the *Elements*—Walras is unequivocal on this point—refers, instead, to 'arbitrary' amounts of initial data.[49] Indeed, a steady-state with $g = \dot{K}_i/K_i$ and constant P_i and R_i, would require that the relative amounts of each $\bar{K}_i$ in Part IV must be identical to the relative amounts of K_i determined in Part V—those consistent with uniform returns—from the very outset. Positive net savings, with rate of growth g (ignoring land), would, as a result, leave both R_i and the price of each capital good unchanged.

But this cannot be attributed to Walras. Recall that Part IV of the *Elements* assumes that only consumer goods are produced. With net accumulation in Part V, as resources are transferred to the production of capital goods (with no assumption of identical productivities), the values of R_k (and hence R_i), R_l and R_n vary in the *tâtonnement* which generates uniform returns. Since costs of production are increasing in K_i, the production of new capital goods, in relative amounts that guarantee r, must imply a change in both prices and the relative quantities of *ab ovo* amounts of K_i in Part IV. In sum, there is *no* evidence, either implicit or explicit, of steady-state growth in the *Elements*.

The same can be said, contrary to Hicks (1934), Diewert (1977) and Morishima (1980, 1977, 1964), for static price-level expectations. Admittedly, Walras (1954:310), at one point in Lesson 28, does recognize that expected and actual prices for capital goods may be different:

> The net income from new capital goods is not known to the same extent as the income from existing capital goods; the net income from new capital goods may prove to be larger or smaller (than expected), in short, it is attended with more risk.

And he also allows for the fact that changes in expectations may change actual values:

> We must remember that the price of capital goods varies not only by reason of past changes but also by reason of expected changes either in gross income or in rates of depreciation and insurance; and that, especially with regard to future changes, expectations differ from in-

> dividual to individual. It follows that numbers of people will sell capital goods which, rightly or wrongly, they fear may be subject to a diminution in net income, and buy with the proceeds other capital goods which they hope, rightly or wrongly, will enjoy an increase in net income. (1954:310–11)

But there is no suggestion that the expected value of future prices is assumed to be constant. There is no suggestion of a tracing out of a sequence of equilibria, where the expected value of certain variables at $t + 1$, e.g., R_i and P_i, depend on actual values at time t.[50] Expectations are simply assumed to 'differ from individual to individual', without any specific assumption about their properties.

The whole issue does not even appear in Lessons 23–25, the formal equations for the theory of capital formation, and for good reason.[51] The 'tickets' device in the *tâtonnement* precludes a difference between expected and actual values. If individuals, for example, are 'wrong' about their estimates of P_j, they simply need to re-issue their 'ticket' demands for Q_j until equilibrium prices are found. On the supply side, in particular, Walras (1954:292) assumes that entrepreneurs 'know their costs of production' as well as 'their selling prices'; so that 'they know in advance whether they will make a profit or loss.' For goods in which there are economic losses, producers 'retire all the *tickets* they have issued to represent' the production of that good. Prices and the 'pledge' amounts of each output adjust until equilibrium is established, with certainty. The assumption of static price-level expectations is certainly not part of the argument; in fact, it is not to be found in the *Elements* at all.

3.4d *Part V as a stationary long-run equilibrium*

If not steady-state growth or static price-level expectations, where does this leave Walras? The evidence strongly indicates that Walras's intention in Part V—in effect, it is the only remaining option, the only interpretation left after steady-state growth and static price-level expectations are ruled out—is to set out the equations for a *stationary* long-run equilibrium. Positive net savings is simply required to 'move' the system, with changes in the relative amounts of each K_i (and hence K_i), from the long-run equilibrium in Part IV of the *Elements*, with input-specific returns (on $\bar{\mathrm{K}}_i$), to the long-run system in Part V, where uniform returns are earned on all capital goods K_i.

As we have seen, *all* of the relevant equations in Lessons 23–25, in fact, designate a stationary long-run equilibrium. The definitions, for example, of the per-unit price of 'perpetual net income'

$$P_E = (1 + 1/(1 + r) + 1/(1 + r)^2 + 1/(1 + r)^3 + \ldots) - 1 = 1/r \tag{3.4.21}$$

and uniform returns

$$r = (R_i + \delta_i P_i)/P_i = R_i/P_i \tag{3.4.22}$$

both give constant (stationary) relative prices over time. Moreover, Walras (1954:289–90) specifies an equality between the rate of interest (the bank rate) and the 'rate of net income' as a condition of equilibrium. In terms of equation (3.1.9), this implies that γ_i across different capital goods is zero. There are no capital gains or losses. Since steady-state growth is not assumed, equation (3.4.22) must read as a stationary long-run outcome, namely the case, 'under certain normal and ideal conditions', where the 'prices of capital goods are rigorously proportional to their net incomes.'

The question, however, is why does Walras initially restrict the Part V analysis to positive net savings, while at the same time holding the amounts of (**l**, **n**, **k**) constant, as given from Part IV, so that new capital goods enter the system only 'later', or in a 'period subsequent to the one under consideration'? The answer lies in the nature of the *tâtonnement* for capital formation itself.

For Walras, the essence of the *tâtonnement* is to describe the process of 'convergence' to equilibrium, leaving the initial data unchanged.[52] Since he begins in Part IV with (**l**, **n**, **k**), in arbitrary amounts, new capital good production, by design, cannot affect the pre-existing stock of $\bar{K}_i$ *until* a uniform rate of return is established in the *tâtonnement* for Part V. This is why K_i enter the economy only 'later'. To trace out the effects of new capital goods as net additions, which enter 'now', with various rates of return, would imply an exercise in *short-run* comparative-statics for Walras; an exercise kept rigorously distinct from the *tâtonnement* throughout the *Elements*.

With $\bar{K}_i$ 'intact', the only way to change the relative amounts of each capital good, in Part V, is with positive net savings. The process of depreciating the pre-existing stock of $\bar{K}_i$ and re-investing these 'funds' in different capital goods (with gross savings alone) is not allowed, since it would necessarily change $\bar{K}_i$.[53] In fact, since input coefficients are fixed and given until Part VII of the *Elements*, there is no way in Part V to *formally analyze* the effect of how K_i, with required variations in a_l, a_n and a_k, changes (**l**, **n**, **k**).

Walras, in other words, unlike Clark (1965 [1900]), does not formally begin—so strong is the idea that *tâtonnement* must leave the total inputs from Part IV unchanged while analyzing displacements around the long-run equilibrium of Part V—with K_i (old plus new additions) endogenous. Instead, the point of 'tickets' in the *tâtonnement* for capital formation, while leaving $\bar{K}_i$ constant, is to find the *equilibrium* amounts of newly produced capital goods (those consistent with uniform rates of return across *all* capital goods, new and pre-existing) first or from the very outset.[54] Once accomplished, the stationary values of R_i, P_i, K_i and r are determined.

Indeed, in Lesson 28, Walras (1954:308) writes the long-run system in reduced form as

$$S = ((R_1 - \delta_1 P_1)K_1 + (R_2 - \delta_2 P_2)K_2 + (R_3 - \delta_3 P_3)K_3 + \ldots)/r \qquad (3.4.23)$$

and

$$P_i = (R_i - \delta_i P_i)/r \qquad (3.4.24)$$

where (3.4.23) 'determines r; and the remaining i equations determine K_1, K_2, K_3, . . .' and remarks:

> If we suppose that the old fixed capital goods of the types k are already found in the economy in quantities K_1, K_2, K_3, . . . respectively and that their gross and net incomes are paid for at prices determined by the system of production equations and by the rates of depreciation and insurance, it is not at all certain that the amount of savings S will be adequate for the manufacture of new fixed capital goods proper in just such quantities as will satisfy the last i equations of the above system. In an economy like the one we have imagined, which establishes itself *ab ovo*, it is probable that there will be no equality of rates of net income.

But, Walras adds:

> On the other hand, in an economy in normal operation which has only to maintain itself in equilibrium, we may suppose the last i equations to be satisfied.

In brief, *once long-run equilibrium is obtained (i.e., when equations (3.4.24) are satisfied), prices are constant over time, with zero net savings.*[55] The system is stationary. Although Walras fails to make the condition $S(\cdot) = 0$ explicit, he all but states it; it is the only meaning that can be attached to an 'economy in normal operation', one which 'only has to maintain itself in equilibrium'. Without zero net savings, rates of return—as Walras (1954:308) recognizes—cannot be uniform across different capital goods and, with variable prices over time, the value of P_E would no longer be a measure of 'perpetual net income'. Equilibrium, as Past V defines it, would simply not be 'maintained'; further accumulation, once r is determined, contradicts the way in which the system of long-run equations is set out.

If this view is correct, the Part V system contained in the *Elements* is a direct analog to the standard equations for long-run equilibrium in Sections 3.1 and 3.2 above, with the important proviso, for Walras, that the value of aggregate capital is an unknown and not a datum of the analysis. Equation (3.2.16) is replaced by

$$K_i = \mathbf{K}_i + \bar{K}_i \tag{3.4.25}$$

with uniform returns on each K_i; the amount of savings

$$\mathbf{S} = q_E = \delta_i P_i K_i \tag{3.4.26}$$

summed over i, simply covers depreciation; and the value of capital, given by

$$K = P_1K_1 + P_2K_2 + P_3K_3 + \ldots \tag{3.4.27}$$

must be endogenous to equilibrium prices and quantities. In the case, for simplicity, in which there are only two produced consumer goods, and two capital goods, the complete system becomes: (3.1.1)–(3.1.4), (3.1.6), (3.1.10), (3.1.11), (3.2.11), and (3.4.25)–(3.4.27), or twenty-eight independent equations in the same number of unknowns [with $P_a \equiv 1$]: Q_j, K_i, K_{ij}, N_j, L_j, N_i, L_i, R_i, R, W, P_b, P_i, K_i, S, K and r. Once equilibrium is established, $\dot{K}_i$ equals zero so that the difference between K_i and $\bar{K}_i$ must simply be the total or accumulated amounts of new capital goods K_i produced, or $\mathbf{K}_i$ in equilibrium.

All of this is quite important. It means, contrary to Garegnani (1976), Robinson (1961), Schumpeter (1954), Morishima (1977), etc., that Walras set out a fully coherent and consistent theory of long-run equilibrium, without the auxiliary assumptions of steady state growth or static price-level expectations. Long-run equilibrium for Walras is *stationary*.

3.5 Conclusion

It is clear that Walras does not take the value of service prices, determined in Part IV of the *Elements*, as fixed in the determination of long-run equilibrium in Part V. Contrary to standard accounts, neither steady-state growth nor static price-level expectations is assumed. The Part V system, instead, represents a stationary equilibrium. The prices of capital goods (as well as their rentals and the quantities produced) adjust to their stationary values in the *tâtonnement* for capital formation. Walras does not err here. Indeed, the only difficulty with Walras's exposition is a failure to make the condition of zero net savings properly explicit. Nevertheless, all of his equations indicate that this condition is, in fact, part and parcel of his theory of long-run equilibrium. The source of this ambiguity—what makes Walras's system *appear* to be wholly non-stationary—is contained in the nature of the *tâtonnement* for capital formation itself; a procedure which requires, for Walras, positive net savings to 'move' the economy from the long-run system in Part IV to the one contained in Part V.

Notes to Chapter 3

1 Semi-stationary or steady-state growth is an example. For the case of a stationary long-run equilibrium, depreciation allowances leave the structure of the capital stock unchanged, i.e., investment is such that the relative amounts of each capital good remain constant.

2 Ignore, however, in a decidedly non-Austrian way, the lapse of time between the production of each capital good as output and its application as input. Cf. Kennedy (1968) and Hicks (1973). An alternative notation is to define Q_j as an index over both consumer and capital goods, where the latter also appear as inputs in the function F_j. The text allows for a clear distinction between capital as a produced input and non-produced or 'original' factors of production, since capital goods do not, so to speak, produce themselves. Cf. Burmeister (1980: 14–5) and Petri (1978), and Lange's (1936) two-sector, 'circulating' capital model of axes (with labour) to produce axes and axes, in turn, to produce wood. The functions F_j and G_i are assumed to be non-identical and neoclassical in all respects, i.e., they satisfy all Inada-conditions: complete substitutability, homogeneity of degree one, diminishing returns throughout to each input taken separately, and so on.

3 The rate of depreciation δ is given exogenously and assumed to be identical for each K_i. In discrete form, $\dot{K}_i$ becomes $\Delta K_i \cdot \Delta t$ while δK_i becomes $[(1 - \delta) \cdot \Delta t]K_i$. Nothing in the following argument changes, in terms of the number of equations less the number of unknowns, if capital goods also appear as inputs to produce themselves.

4 Lange (1936:161n3) notes the possibility, without a substantive change in results, of writing the supply of labour as (say) an increasing function of net output.

5 Equation (3.1.8) is the maximized value, over the choice of $(\mathbf{k}, \mathbf{n}, \mathbf{l})$, of (3.1.5). See Hahn (1982).

6 Assume (for now) myopic perfect foresight so that the right-hand time derivative of $P_i(t)$ is known with certainty.

7 The steady-state case is more complicated than suggested here since the value of γ_i depends on the choice of *numéraire*. Regardless of that choice, however, real returns are equalized. It is understood, in the form of equation (3.1.10), that depreciation allowances accrue at the rate r. Since there is no 'money' in the system—and no way to hold such allowances over time—it is necessary to assume, as a condition of stationary equilibrium, that δK_i keeps each 'machine' physically intact. As long as the system is *in* long-run equilibrium, there is no incentive to transfer depreciation allowances, in terms of *numéraire*, from one capital good to another. The value $(R_i - \delta P_i)/r$ thus reads as perpetual net income. See Kuenne (1963:205–6, 220–1) and Hirshleifer (1970).

8 Equations (3.1.1)–(3.1.4) guarantee that the supply and demand for inputs are equal. Equilibrium implies that the $ED_b = 0$. Production and exchange are assumed to take place only at equilibrium values. Since the functions F_j are non-idential, or employ different combinations of inputs, the supply of Q_a is increasing in P_a.

9 An equality, of course, between the number of equations and unknowns is not sufficient to guarantee a solution, much less a unique solution. The text ignores the 'existence' problem.

10 Clark's (1965 [1899]) approach is to add an equation which expresses an equality between the supply and demand for the aggregate value of capital (K); erroneously taking K as exogenous, to form an analog to equations (3.1.4) or the resource constraints for the homogeneous stocks of labour and land.

11 This procedure needs some explanation. Walras does not allow for substitution in (3.1.1) until Part VII of the *Elements*. The input coefficients in the Part IV price equations

$$P_j = R_i k_{ij} + W n_j + R l_j$$

are fixed and known *a priori*. However, Walras (1954:240) does remark, in Part IV, that k_{ij}, n_j and l_j are *not* actually 'determined in this way, for in the making of a product, it is possible to employ more or less of some productive services, say land-services, provided that correspondingly less or more of other productive services, say labour or capital services, are employed. The respective quantities of each of the productive services which thus enter into the making of a single unit of each of the products are determined, along with the prices of the productive services, by the condition that the cost of production of the products be a minimum. We shall later express this condition by a system of as many equations as there are coefficients of production to be determined.' The equations derived 'later', i.e., in Part VII, are identical to (3.2.1). As Walras (1954:385) observes: 'In a state of equilibrium, when cost production and selling price are equal, the prices of the services are proportional to their marginal productivities, i.e., to the partial derivatives of the production function.' With prices parametric, note that (3.2.1) contains eight equations in eight unknown input coefficients. Although there is some contextual significance to Walras's use of fixed coefficients, introducing substitution here, in the characterization of the Part IV equations, does not substantively distort the reading of the *Elements*. See Sections 3.3 and 3.4 below. On whether or not Walras was aware of the importance of homogeneity of degree one in product exhaustion, cf. Stigler (1961), Jaffé's 'Translator's Notes' to Walras (1954:549–53) and Jaffé (1983 [1964]:176–212).

12 As Walras constructs the argument, there is no intertemporal connection whatsoever in Part IV of the *Elements*: capital goods are not produced; there is no market for ownership claims to capital; no one saves; and there is no rate of time preference. See Section 3.3 below.

13 In Part IV, Walras (1954:243) assumes m consumer goods and a heterogeneous list of n different inputs. See Section 3.3 below. It should be clear that specifying additional inputs does not alter the system's dimension (i.e., the number of equations less the number of unknowns). This is not strictly the case though with each consumer good that is added, since each Q_j also requires something extra, notably a market-clearing condition. As is well-known, a system with (say) two consumer goods—unless both are produced with identical input coefficients, thus allowing for consistent aggregation—will be qualitatively different from a one-good model. None of this matters, however, in terms of the key properties (e.g., zero economic profits, income equals consumption, no uniform rate of return or prices for capital goods, etc.) that characterize Walras's long-run equilibrium in Part IV.

14 Equations (3.2.4) and (3.2.6) are exactly those used by Walras (1954:384–5) in the discussion of marginal productivity theory, Lesson 36 of the *Elements*.

15 Note the ambiguity in the meaning of a 'new' capital good; the beginning of a problem, in Walras, in distinguishing between capital goods as produced as opposed to 'original' inputs. See Section 3.4 below.

16 The price equations (3.2.9) are redundant to first-order conditions. Walras assumes that capital goods also appear as inputs in (3.1.2).

See Section 3.4 below. Adding these to the equations in the text would not alter its dimension, i.e., with the appropriate change in resource constraints, each capital good, used to produce other capital goods, adds as many first-order conditions as unknowns.

17 Note Collard (1973:474): 'Walras's theory is then appropriate only to capital accumulation at a single moment in time when relative prices are expected to be constant *forever.* Once new capital goods are available to assist in production a whole new set of relative prices is expected to persist *forever* and a new set of decisions about accumulation is taken; and so on. The movement is not necessarily steady but any change is taken, for that moment, to be permanent. At each round the outcome is expected to be a new stationary state.'

18 Note Diewert (1977:74): 'In fact, in some ways, Walras's model goes beyond that of Hicks since Walras carefully relates the purchase (or stock) price of a durable to its rental (or flow) price (under the additional simplifying assumption of static expectations) whereas Hicks works entirely in terms of flow prices.'

19 Explicit reference, however, is made to Hicks (1965:58–75) where, as a direct line from Walras [Morishima (1980:554n4)], static price-level expectations *are* assumed. A similar argument is made in Morishima (1977:79–81), although here, in the formal representation of Walras, Morishima (1977:82) simply assumes that capital goods cannot be traded, so that the capital gains or losses in equation (3.1.9) are no longer relevant. Walras is later criticized [Morishima (1977:202)] for ignoring such effects. Morishima (1964) analyzes the stability of a Walrasian system with static price-level expectations.

20 This is exactly the way in which Kuenne (1963:202–19) sets out the Walrasian system.

21 Montgomery is careful to point out that steady-state growth is, in fact, a *re*-interpretation of Walras, but he also claims: 'Walras's recognition of the existence of a problem of changing expectations and his attempts to avoid its difficulties lead to the hope that he might concur in this interpretation.'

22 Garegnani (1976), Robinson (1961) and Petri (1978), as discussed in Chapter 1 above, basically share this view. In Jaffé's later work [e.g., the 'Translator's Notes' to the *Elements* (1954:526) and Jaffé (1967 [1983]:228)], he does recognize that economic profits in the Part IV system are zero. He nowhere qualifies, however, his earlier statement that (as a result) Part IV of the *Elements* does not correspond to a Marshallian short-run.

23 Walras (1954:224–5). Note Walras (1954:240): 'We are neglecting another matter, namely that of the distinction between fixed and variable costs in business. But since we are assuming that entrepreneurs make neither profit nor loss, we may as well assume that they are also manufacturing equal quantities of products, in which case costs of all kinds may be considered as variable.'

24 Some of Walras's notation is changed for convenience.

25 With an aggregate budget constraint, one of the production equations is redundant. In equilibrium the total demand for inputs (equal to their 'effective offers') must equal the total amounts of (**l**, **n**, **k**) as data.

26 Note Walras (1954:225): 'In a state of equilibrium in production, entrepreneurs make neither profit nor loss. They make their living not as entrepreneurs, but as landowners, labourers or capitalists in their own or other businesses.'

27 See Chapter 2 above.

28 Even in terms of this simple argument, there is a serious inconsistency that is worth noting. In Part II of the *Elements* Walras writes utility functions as additively separable across goods. In a multi-market context this implies, with a fall in a particular price, that the prices of all other goods either rise or fall together, depending on the elasticity of demand for the particular good in question, i.e., all other prices move in the same direction. See Hicks (1956) and Chapter 2 above. In other words, the 'compensating' moves in prices, some rising, some falling, that Walras relies on as part of the intuition for the 'convergence' equilibrium, are simply not compatible with the form of the utility functions that he assumes.

29 Walras (1954:247), indeed, sees this process, at least in some sense, as the way in which actual markets work: 'This is precisely the sort of groping which takes place in the products market under conditions of free competition, as entrepreneurs increase or decrease their output according as they make profits or losses.' From the Preface to the fourth edition [Walras (1954:44)]: 'The upward and downward movement of market prices in conjunction with the effective flow of entrepreneurs from enterprises showing a loss to enterprises showing a profit is purely and simply a method of groping towards a solution of the equations involved in these problems.'

Note Walras (1954:255): 'It follows from the proof we have just given in Lesson 21 that free competition provides a practical solution to the equations set out in Lesson 20, since free competition consists, on the one hand, in allowing entrepreneurs freedom to expand output in case of a profit and restrict output in case of a loss; and, on the other hand, in allowing land-owners, workers and capitalists, as well as entrepreneurs, freedom to buy and sell services and products by bidding against one another.'

30 Walras also includes a technologically given '*insurance premium*', which is ignored here for simplicity.

31 The possibility of 'eating into one's capital' is also acknowledged. See Walras (1954:273).

32 The first three editions simply postulate an 'empirical' or *ad hoc* savings function with a constant average propensity to save. See Jaffé (1942:43–4).

33 Walras assumes that utility is additively separable across goods throughout, so that equation (3.4.7) expresses the familiar condition that optimizing individuals will allocate expenditures over goods and perpetual net income until the ratio of marginal utilities to prices is equal.

34 Note Walras (1954:269–70): 'We have all the elements necessary for the solution of our problem. New capital goods are exchanged against the excess of income over consumption; and the condition of equality beween the value of the new capital goods and the value of the excess gives us the equation required for the determination of the rate of net income and consequently for the determination of the prices of new capital goods. Moreover, new capital goods are products; and the condition of the equality between their selling price and their cost of production gives us the equations necessary for the determination of the quantities manufactured.'

35 See Section 3.4c below.

36 Walras (1954:287–8). In fact, as with most of the functional relationships in the fouth edition of the *Elements*, Walras considers $S(\cdot)$ to be multi-valued, i.e., with both increasing and decreasing ranges. Except for Part II of the *Elements*, however, Walras reads all resulting equilibria, the form of $S(\cdot)$ notwithstanding, as *unique*. See Chapter 2 above. The issue is ignored here, without consequence.

37 Note Walras (1954:289): 'This particular groping which we have just described actually takes place in the stock exchange, which is the market for new capital goods, where the prices of these goods rise (or fall) through a fall (or rise) in the rate of net income, according as the demand for new capital in terms of *numéraire* is greater (or less) than the supply.' Note that the right-hand side of equation (3.4.15) simply expresses the price of each capital good.

38 Along with the effect of variable input coefficients on service prices, there is also an output price-effect, i.e., a change in the relative prices of consumer goods affects S, which Walras (1954:288) largely ignores.

39 Again, Walras (1954:288) also allows for the prices of consumer goods to fall, but the precise effects on savings are not set out.

40 Walras (1954:291) explicitly draws this effect from Lesson 22, Part IV of the *Elements*, i.e., 'The Law of Variation of Prices of Products and Services' [Walras (1954:260)]: '*If, all other things equal, the quantity of one of these products or services in the hands of one or more holders increases or decreases, the price of this product or service will decrease or increase.*'

It must be admitted, however, that the bulk of the argument in Lesson 25 appears to rely, mostly, on increases in the cost of production, the right-hand side of equation (3.4.17), to restore an equality between P_i and P_i^*. The connection between own-service prices and output is muddled in Lesson 25, owing to a number of technical changes that Walras made after the second edition of the *Elements*. See Jaffé's (1954:594) collation notes. The latter make it clear that in the second edition a fall, say, in R_1 [i.e., the left-hand side of (3.4.17)], as a result of an increase in K_1, is equally prominent with a change in costs of production in the *tâtonnement* for capital formation. In light of Lesson 22, there can be little doubt that Walras had this mechanism in mind throughout all the editions of the *Elements*.

Indeed, in the informal discussion, the adjustment in selling prices is predominant in the 'move' to long-run equilibrium, e.g., Walras (1954:271): 'Capital goods proper are artificial capital goods; they are products and their prices are subject to the law of cost of production. If their selling price is greater than their cost of production, the quantity produced will increase and their selling price will fall; if their selling price is lower than their cost of production the quantity produced will diminish and their selling price will rise. In equilibrium their selling price and their costs of production are equal.' See Hollander (1982), in this regard, for the undeniable connection between Walras and the British Classical tradition, including Ricardo.

41 Note Walras (1954:294): 'This is precisely the sort of groping which actually takes place of its own accord in the real market for products under conditions of free competition, when entrepreneurs manufacturing new capital goods, like the entrepreneurs manufacturing consumers' goods, increase or decrease their output according as they make profits or incur losses.'

Walras completely misses the point that in the 'convergence' to equilibrium in Part V, *nominal* rates of return across different capital goods, with perfect foresight, will be equalized. See equation (3.1.9) and Appendix C below.

42 Walras (1954:293) also acknowledges the fact that some capital goods may not be produced at all, i.e., that $\dot{K}_i = 0$, for some i.

43 See Section 3.2 above.

44 This is not to suggest that Walras was unprepared, for certain analytical purposes (in addition to the preliminary system in Lesson 23), to hold Part IV variables constant. The graphs, for example, of the amount of savings and the demand for new capital goods in Lesson 28, with r as the independent variable, necessarily do so. Service prices are listed as parameters. In Appendix I of the *Elements* (the 'Geometrical Theory of the Determination of Prices'), Walras simply neglects—again in a largely graphical treatment—'any disturbances of the assumed equilibrium in exchange and production which may result from the special adjustments towards equilibrium in capital formation.' None of this, however, negates the importance of Lesson 24.

45 Walras (1954:309).

46 Walras (1954:260–62), in fact, graphs the demand for each capital good, in terms of its relative quantity, as inversely related to its own-service price.

47 Recall, for Walras, that only rentals and not prices are determined for each K_i in Part IV. With the 'rate of net income', the price of land is determined by equation (3.2.16).

There is one point, in Part VII of the *Elements*, where the prices of capital goods do remain constant. The context, however, is not one of changing the relative amounts of each K_i, as a 'move' to long-run equilibrium, but a comparative-statics exercise designed to analyze the effect of an exogenous increase in capital. With land as a fixed input, Walras (1954:390–91) assumes, for simplicity, a proportionate increase in each capital good such that all outputs double, leaving—with an increase in rent and, by assumption, a comparable fall in the marginal product of capital—costs of production 'approximately the same'. The result, along Classical lines, is a fall in the 'rate of net income', since R_k falls with a larger capital stock, while P_k remains unchanged.

In general, Walras saw the prices of capital goods as 'quite variable', e.g., Walras (1954:311): 'Of the three systems of equations relating to the market for capital goods, there is one that is particularly important, namely, the system pertaining to capital goods proper. This is so, because the identity of gross and net income from land exempts the price of land from two causes of variation, viz. changes in the rate of depreciation and changes in the premium rate of insurance. Personal faculties are neither bought nor sold where slavery is inadmissible. Only in the case of capital goods proper are the gross incomes, depreciation rates and premium rates of insurance all subject to change, so that their prices are quite variable and they are constantly being bought and sold for speculation.'

48 Note Walras, in a letter to Böhm-Bawerk [Jaffé (1983:32)], dated 5 May 1889: 'You deduce the rate of interest from the difference in value between present and future goods. For my part, I think the difference in value between present and future goods cannot be determined directly, for this difference can only be deduced from the rate of interest which is determined directly by the ratio of the total yield from new capitals to the volume of savings.' Wicksell shared a similar attitude, i.e., that the production-side must first determine r before the ratio of constant relative prices over time could be known. See Chapter 4 below.

49 See Walras (1954:242, 282, 308).

50 Cf. Jaffé (1980) and Morishima (1980). Morishima (1980: 554n4) suggests that the fifty pages which constitute Part VII of the *Elements* carry out this sort of 'sequence analysis'. But there is no evidence for this. The only section of Part VII which *may* come close are the five pages in Lesson 36 [Walras (1954:388–92)] on the 'Law of General Price Movements in a Progressive Economy'. The argument here, however, concerns a comparative-statics exercise of the effect of a larger capital stock on prices and incomes. There is no tracing out of a sequence of equilibria where expected and actual prices vary.

51 Walras (1954:310–1), in fact, relegates the role of 'speculation' and 'expectations' to a study in 'applied economics'. Note Jaffé (1942:48n9): 'Though Walras failed to incorporate the concept of expectations in his theoretical structure, he recognized its importance as an afterthought.'

52 See Chapter 2 above.

53 At one point, Walras (1954:46) explicitly allows for this sort of process: 'Society realizes a gain in utility by withdrawing capital from an employment where it yields a lower interest in order to invest it in an employment where it yields a higher rate of interest.'

Note that with a rate of depreciation, some new capital goods must enter 'now' to keep the relative amounts of $\bar{K}_i$ unchanged as they depreciate with use; again, underscoring the nature of the *tâtonnement* with constant data.

54 See Section 2.4 above.

55 Walras (1954:273–4) is quite clear on the distinction between 'eating into one's capital', adding to the 'supply of new capital goods', and the case where an 'individual holds constant the quantity of the capital goods proper he owns, without increase or decrease.'

Chapter 4

Stationary equilibrium and Wicksell's 'missing equation'

Recall from Chapter 1 above, that standard interpretations argue that Wicksell closes his various systems of equations for long-run equilibrium by taking the aggregate value of the stock of capital as a datum. Since the latter is clearly an endogenous variable, Wicksell is thus portrayed as being caught in a 'circular reasoning', erroneously attempting to determine equilibrium prices with a prior knowledge of the prices used to define the measure of capital. In this chapter, I examine Wicksell's long-run systems in detail and show that the standard view of his theory is incorrect.

Section 4.1 below gives a synopsis of the series of technical relationships that characterize Austrian capital theory. Section 4.2 examines *Value, Capital and Rent* [hereafter *VCR*], drawing, where appropriate, from relevant material in the first volume of Wicksell's *Lectures on Political Economy* [or simply, the *Lectures*]. With regard to the definition of a stationary equilibrium, exactly the same arguments are used by Wicksell in both works.

Specifically, Section 4.2a sets out the variety of conditions contained, for Wicksell, in the idea of a stationary economy, e.g., constant income, consumption, and relative prices over time, zero net savings, and so on. For the most part, the systems in *VCR* assume a single 'circulating' capital good, corn, used with labour and land, to produce corn. The stock of each 'fixed' or durable capital good is held constant, generating input-specific rentals, with a uniform rate of return earned only on 'circulating' capital or the 'advance' of corn over the period of production. This is what Wicksell later termed a 'provisional' long-run equilibrium.

Section 4.2b examines this corn model, the analytical core of

VCR, while 4.2c sets out the bulk of the case for Wicksell's recognition of the need for a 'theory of accumulation', including an explicitly stated equality between the rate of time preference (ρ) and the rate of return to capital (r) to close his stationary long-run systems. Part of the explanation for the appearance of 'a one-sided productivity model' in Wicksell, in fact, stems from his claim that stationary conditions *per se* imply a zero rate of time preference. The productivity-side of the model is used by Wicksell not only to explain the rate of return on capital, but to guarantee and determine a positive ρ.

In Section 4.2d, Wicksell's two-consumption good, two-capital good model is discussed. It is important to note that at the only point in *VCR* where the presence of more than one 'circulating' capital good (corn and linen) is allowed, Wicksell is clear that the value of aggregate capital cannot be taken as exogenous, i.e., that its measure must depend on equilibrium prices.

Section 4.3 is devoted entirely to the *Lectures*. Section 4.3a, in particular, spells out the way in which the arguments in *VCR* are amended in Wicksell's later work; notably, the idea of an 'average period of production' is dropped. Section 4.3b examines the celebrated wine model. Contrary to standard accounts in the secondary literature, Wicksell, in fact, closes this system (excluding land) with an assumption regarding the equilibrium price of matured wine across successive vintages. The value of capital (K) is explicitly written as an endogenous variable. Surprisingly, the view that Wicksell's wine models are open—with, as a consequence, Wicksell choosing the theoretically indefensible position of arbitrarily fixing K—is held with virtual unanimity, e.g., Lutz (1968), Negishi (1982a), Samuelson (1982), Sandelin (1980, 1982) and Stigler (1961). Although Samuelson and Negishi go on to suggest that it was Wicksell's (apparently halfhearted) 'intention' to close these systems with a 'theory of savings behaviour', my interpretation shows that not only does Wicksell's standard wine model not take the value of aggregate capital as exogenous, but that Wicksell, throughout *VCR* and the *Lectures*, both recognized the need and set out the required conditions, in terms of savings behaviour or intertemporal preferences, to close all of his systems of equations properly.

Section 4.3c considers the two-consumption good, two-capital good model contained in the *Lectures*. A key difference, from the corresponding system in *VCR*, where K is the aggregate value of corn and linen, is the measure of capital as a Sraffa-reduction to 'dated' quantities of labour and land inputs. The qualification that the value of aggregate capital is endogenous to both equilibrium prices and a savings market is prominent. Section 4.4 concludes.

4.1 A review of Austrian fundamentals[1]

An essential aspect of the Böhm-Bawerk/Wicksell theory of production is the idea of capital as a bridge, across time, between the application of inputs and the realization of output. The length of the 'period of production' (τ) provides the basis for a measure of capital relative to total wage payments. Longer or more 'roundabout' production processes require relatively more capital as an 'advance' to per-unit labour, before output is realized, and deliver both increases in labour-productivity and larger amounts of net output. Time thus becomes a measure of capital-intensity. The more capital-intensive is the production process, the larger the share of net output that accrues to labour as opposed to capital.

Consider first the well-known case of point-input/point-output, e.g., Wicksell's natural-ageing processes, such as the maturing of wine or the growth of a tree. Inputs occur only at time zero so that labour is simply used, so to speak, to store grape-juice or plant seedlings. Represent net output, the amount (say) of wine realized at the 'end' of a production process, by

$$q \equiv Q/N = f(\tau) \tag{4.1.1}$$

where q depends on the amount of labour (N) and the period of production, for $f'(\tau) > 0$, $f''(\tau) < 0$ and $f'(\infty) = 0$. Let W represent the wage rate and assume constant returns to scale. In equilibrium, the value of net output, with continuous compounding, is

$$q \equiv Q/N = We^{r\tau} \tag{4.1.2}$$

where

$$r = \partial q/\partial\tau / q = f'(\tau)/f(\tau) \tag{4.1.3}$$

gives the 'marginal productivity of time', for r the rate of return. There are no durable capital goods—the meaning of point-output—or all capital is 'circulating', fully depreciating with use.

The value of capital is

$$k \equiv K/N = W \int_0^{\tau} e^{rt} dt = W(e^{r\tau} - 1)/r \qquad (4.1.4)$$

where combining (4.1.2)–(4.1.4), results in

$$k = (q - W)/r \qquad (4.1.5)$$

or 'product exhaustion' as a condition of long-run equilibrium. If the system is stationary, there can be no net accumulation and no dis-saving, i.e., for the stock of capital to remain intact, capitalist consumption, out of net output, must be given by rk. Fixing $N = \bar{N}$, finally, gives [with (4.1.1)–(4.1.4) and constant returns to scale] five independent equations in six unknowns: Q, K, N, τ, W and r. There is a 'missing equation'.

To close the system, add an expression representing savings behaviour; a condition defining the marginal rate of time preference across consumption bundles over time. Assume two periods and write

$$U = U(c_t, c_{t+1}) \qquad (4.1.6)$$

as the utility from the present and future consumption of wine. Let the rate of time preference be a positive constant. With a given amount of labour (i.e., with the growth rate of the labour force set equal to zero), all capital 'circulating', and zero net savings, each unit of wine, as capital-*cum*-wage payment, is simply used and replaced, one-for-one, with production. Maximizing (4.1.6), with the price of future wine discounted to the present, i.e., with constant-value prices, gives

$$1 + \rho_t = \partial U/\partial c_t / \partial U/\partial c_{t+1} = 1 + r_t = p_t/p_{t+1} \qquad (4.1.7)$$

as the 'missing equation' (in discrete form), for ρ_t a given rate of time preference, and $r_t = (p_t - p_{t+1})/p_{t+1}$ the wine own-rate of interest [see Sections 4.2c and 4.3b below]. Variations in r_t define a savings function with $\rho = r$, as a condition of equilibrium (in continuous terms), implying that $dc/dt = \dot{c} = 0$.

Extending the point-input/point-output case to allow for an exogenous growth rate of the labour force is straightforward. Replace $N = \bar{N}$ with

$$N_\tau = N_0 e^{gt} \qquad (4.1.8)$$

for $g = \dot{N}/N$, a constant. The value of capital now becomes

$$k \equiv K/N_\tau = 1/N_\tau \int_0^\tau W e^{rt} N_t e^{-gt} dt = (We^{(r-g)\tau} - 1)/(r - g) \qquad (4.1.9)$$

for $r \neq g$. Note that when $r = g$ the value of the integral in (4.1.9) is simply

$$k \equiv K/N_\tau = Wt. \qquad (4.1.10)$$

A steady-state implies that the value of capital, even with continuous compounding, is given by Wt, or the size of the 'subsistence fund' as an 'advance' to labour.

All this can be approached from a slightly different direction, one that is especially prominent in *VCR*. Consider the case of continuous-input/point-output for a stationary economy. Assume a single, produced output with a period of production τ—say, the planting and continuous cultivation of corn. Labour is applied uniformly over time, indexed by t, and capital, as corn, is 'advanced' as a wage payment. As a condition of stationary equilibrium, the production of corn must be uniformly distributed across all levels of maturity, i.e., with the first realization of output, there are, in discrete terms, $\tau - 1$ processes simultaneously in place, each at a different stage from completion, each with a period of production τ. Net output is a continuous or permanent flow of corn over time, or a constant amount available for consumption at each t.

The value of the capital stock, integrating over all maturities, is given by

$$\mathit{k} \equiv \mathit{K}/N = \int_0^\tau [W/r\tau]\,(e^{rt} - 1)dt = [W/r^2\tau]\,(e^{r\tau} - 1 - r\tau) \qquad (4.1.11)$$

where the standard approximation to (4.1.11) is

$$\mathit{k} \cong k \equiv K/N = W/r^2\tau\,(1 + r\tau + \tfrac{1}{2}(r\tau)^2 - 1 - r\tau) = \tfrac{1}{2}W\tau \qquad (4.1.12)$$

for $T = \frac{1}{2}\tau$ as the 'average period of production'. Long-run equilibrium again implies zero economic profits or 'product exhaustion'

$$q = f(\tau) = W + rk \qquad (4.1.13)$$

so that, using equation (4.1.12),

$$f'(\tau) = \tfrac{1}{2}Wr \qquad (4.1.14)$$

defines the 'marginal productivity of time' as an analog to equation (4.1.3).

The idea of an 'average period of production' is in fact quite simple. In Böhm-Bawerk (1959:II [1889]), T is the arithmetic average of uniform applications of labour, weighted by the length of time from the point of input to the point of final output, now realized at $t = 0$. With a uniform distribution of production processes, the number of discrete time-periods is $h = \tau - 1$. Working backward through time, define α_t as the amount of labour, per-unit of output, applied at t; so that α_o is an application of labour at the point closest to, and α_h, the point furthest from, final output. The average period of production is

$$T = \sum_{t=0}^{h} t\alpha_t \Big/ \sum_{t=0}^{h} \alpha_t \qquad (4.1.15)$$

or the per-unit ratio of time-augmented to total labour. As the number of discrete time-periods becomes infinitely large, i.e., as time becomes continuous, $T = \frac{1}{2}\tau$. For a given wage rate, (4.1.15) measures

$$T = W \sum_{t=0}^{h} t\alpha_t \Big/ W \sum_{t=0}^{h} \alpha_t = K/WN \qquad (4.1.16)$$

or the ratio of the amount of capital as a 'subsistence fund'—in corn, as intermediate product, 'locked' in various stages of production—to net output, less the return on capital.

Note, in particular, that the construction of T assumes the use of simple interest. To see this, explicitly write the per-unit production process as interest-augmented wage payments

$$P = W\alpha_0 + W\alpha_1 (1 + r) + W\alpha_2 (1 + 2r) + \ldots = W\sum_{t=0}^{h} \alpha_t (1 + tr) \qquad (4.1.17)$$

for P the per-unit value of output. In a stationary economy

$$Pq = W \sum_{t=0}^{h} \alpha_t (1 + tr) \Big/ \sum_{t=0}^{h} \alpha_t = W + rk \qquad (4.1.18)$$

is the value of per-capita net output. Solving for rk

$$rk = W \left(\sum_{t=0}^{h} \alpha_t (1 + tr) \Big/ \sum_{t=0}^{h} \alpha_t - 1 \right) = rW \left(\sum_{t=0}^{h} t\alpha_t \Big/ \sum_{t=0}^{h} \alpha_t \right) \qquad (4.1.19)$$

gives

$$k \equiv K/N = WT \qquad (4.1.20)$$

regardless of the value of $r > 0$. With simple interest, the value of T (setting savings behaviour aside) is independent of the rate of return on capital.

Now work solely in terms of T, a measure of capital-intensity. Net output per capita

$$q \equiv Q/N = f(T) \qquad (4.1.21)$$

is a function of the average period of production; for $f' > 0$ and $f'' < 0$. Generally, Böhm-Bawerk and Wicksell consider T to be finite, but leave open the possibility that the 'marginal productivity of time' may be driven to zero for some sufficiently large value of the average period. In stationary equilibrium

$$q = W(1 + rT) \qquad (4.1.22)$$

or

$$r = (q - W)/WT \qquad (4.1.23)$$

gives the rate of return on capital as a share of net product, in terms of wages 'advanced' (on average) over t. Maximizing (4.1.23) with respect to T, taking W as parametric, gives

$$r = f'(T)/W \qquad (4.1.24)$$

or

$$W = f(T) - Tf'(T) \qquad (4.1.25)$$

one of which adds an independent equation; with $dr/dT = f''(T)/W < 0$ and $dW/dT = -f''(T) > 0$. The wage rate in (4.1.25) is the difference between net output and the share of net output accruing to capital. An increase in the average period of production raises W and lowers r; makes labour more productive, capital less so.

Equations (4.1.20)–(4.1.22), and (4.1.24) are four in number to solve for five unknowns: q, T, k, W and r. An equation comparable to (4.1.7) closes the system, providing the 'missing equation' for a given ρ. Fixing $N = \bar{N}$, gives Q and K measured in corn.[2]

Note the importance of holding W parametric—the firm's point of view—in the derivation of (4.1.24). Totally differentiating (4.1.20) gives

$$dK/dT = T \cdot (dW/dT) + W = W \qquad (4.1.26)$$

for $N = \bar{N}$ and $dW/dT = 0$. From (4.1.24)

$$r = (1/W) \cdot (dq/dT) = (dT/dK) \cdot (dq/dT) = dq/dK = f'(T) \qquad (4.1.27)$$

so that, in equilibrium, r equals the 'marginal product of time', or the return to capital. For the economy as a whole, however, $dW/dT = (dW/dK) \cdot (dK/dT) > 0$. An increase in the stock of capital is partially 'absorbed' in higher wage payments; or

$$\begin{aligned} r &= (1/W) \cdot (dq/dT) > f'(T) / dK/dT \\ &= (dq/dT) \cdot (TdW/dT + W) = \hat{r} \end{aligned} \qquad (4.1.28)$$

implies that the private rate of return on capital exceeds the social rate or $\hat{r}$. Even in this simple context the 'Price-Wicksell Effect' is already evident.

4.2 *Value, Capital and Rent*

4.2a *The characteristics of a stationary equilibrium*

Wicksell (1934, I [1911]:105) is clear that the theory of production and distribution, contained in both *VCR* and the *Lectures*, applies only to 'the *static* aspect of the problem of equilibrium, i.e., the conditions necessary for the maintenance, or periodic renewal, of a *stationary state of economic relations*'.[3] The assumption of a stationary economy, 'as a fundamental—and simplest—hypothesis', is used by Wicksell to place Böhm-Bawerk's theory of interest (as an 'agio or premium which arises from exchange between present and future goods') in a framework consistent with von Thünen's 'productivity' approach.[4] Given an equilibrium capital stock, variations in the length of the period of production determine r, or the rate of return to capital. But r must also conform to the ratio of marginal utilities of present to future goods, the 'formula' which 'governs the problem of interest in its whole extent'.[5]

The stationary state, for Wicksell, has four related components. First, the stock of 'circulating' capital (e.g., the value of corn, with a turnover of $\frac{1}{2}\tau$), in particular, remains unchanged. As Wicksell (1954:103) indicates:[6]

> It will perhaps be best, if we are to find our way in this rather complicated state of affairs, to base our thinking throughout upon the assumption of a *stationary community*, as the simplest hypothesis. For all the productive factors, and consequently capital too, could then be considered as approximately *constant* magnitudes. Though in this case the forms of the latter change, its total value remains unchanged, since in place of the consumed capital goods new ones of equivalent value enter successively.

The value $K = \frac{1}{2}WN\tau$ is paid as wage-income and replaced with production; the amount—and only the amount—rK is consumed by capitalists, leaving the stock of K intact.[7]

Second, net income, expressed by equation (4.1.22), and the relative exchange value of goods (with r greater than zero) are constant over time.[8] This implies, with a uniform distribution of production processes, that the time-structure of the 'subsistence fund' is always appropriate to the length of the average period of production, i.e., that the flow of total output and the cost of inputs match, at each point in time, leaving net output unchanged. As Wicksell (1954:124–5) notes:

> The entrepreneur carries on not merely one, but several businesses of the same kind at the same time, in all of which the period of production is the same, but which are at different stages of progress, so that the entrepreneur can consequently market finished goods once a month, say, or once a week. The proceeds from these provide him with the necessary money for the next payment of wages.

'This is the same as saying,' he concludes, 'that the half-finished products are produced precisely when consumption and production require them.'

Third, the stock of each durable (or fixed) capital good is constant, or remains intact, by assuming that depreciation allowances are always 're-invested', so that in 'a stationary economy', as Wicksell (1954:162) states it, 'such goods are not produced at all, but are kept in the same good condition.' Wicksell (1954:119), in fact, counts only 'circulating' capital—goods that fully depreciate with use—in the value of K; the measure most relevant in determining W and r:[9]

> Goods of greater durability (such as streets, railways, buildings, etc.) cannot be regarded or treated as capital in the narrower sense, but, once they are there, must be placed, economically speaking, in the same category as landed property itself. In other words, if, in accordance with Böhm-Bawerk's precedent which we ourselves shall follow, all existing capital goods are united in one sum, in order to use this sum as an element in the theoretical determination of the level of interest and wages, it would be misleading to think of the capital value of all railways, buildings, etc., as being included in this sum. This value is rather, like the capital value of landed property itself, to be thought of as a secondary phenomenon which has no influence on the determination of the above magnitudes. The net interest of durable goods, however, is determined, like ground-rent, simply by the value of their useful services (after the cost of repairs has been deducted).

The kind of long-run equilibrium, in other words, that Wicksell has in mind—what he later termed a 'provisional equilibrium'—excludes both net increases in fixed capital, and any adjustments to the structure of capital obtained by reallocating depreciation allowances from one durable good to another. The return to each fixed capital good (as with the given amounts of land and other natural resources) is thus input-specific, having the 'precise character of a rent'. Uniform returns are earned on 'circulating' capital alone; not, in contrast to Walras (1954 [1900]), on all capital goods, as a uniform ratio of net rentals to the supply price for each.

There seem to be two reasons for adopting this procedure. The first and most obvious is that the presence of fixed capital (the case of continuous-output as opposed to point-output) makes the average period of production, equation (4.1.15), all but impossible to define.[10] Moreover, with durable goods, net output per capita [(4.1.21)] can no longer be represented as a simple function of T. These are points which Wicksell later emphasized in the *Lectures*. The analytical core of *VCR*—a single 'circulating' capital good, generating a stationary flow of physically identical net output, with simple interest accruing—avoids these problems.

Durables simply enhance productivity, and are envisaged as exerting the 'same influence as, say, the size of the cultivated area of land'. As such, they are often assumed by Wicksell, for convenience, either not to exist, or to be free goods in given amounts.[11]

The second reason for Wicksell's procedure concerns the length of time required for the adjustment to long-run equilibrium. The argument first appears in the *Lectures* where the complexities associated with a number of different capital goods, with varying durabilities, is more thoroughly recognized. Here, Wicksell argues that many capital goods, particularly highly durable ones, are non-shiftable; that 'machines' are output-specific so that they cannot be moved from one employment to another instantly, if at all. To generate a uniform rate of return over all capital goods (for full adjustment) requires a 'slow' process of reallocating depreciation allowances, across different capital goods, changing their relative amounts and supply prices. To avoid the problem with the length of time necessary for such an adjustment—to avoid extending 'our observations over periods of time in which centuries are mere episodes'—Wicksell (1934, I:166) chooses to limit the long-run analysis to a 'shorter period'. Although he recognizes that 'there is always a *tendency*, perhaps very incompletely realized,' for changes in the relative amounts of fixed capital to generate uniform returns, the analysis is restricted to non-durables or goods of 'limited' durability. As Wicksell (1934, I:186) explains:

> Nearly all long-term capital investments, nearly all so-called fixed capital (houses, buildings, durable machinery, etc.) are, economically speaking, on the border line between capital in the strict sense and land. We have already said that the operation of the laws of capital depends upon the assumption of a constant adjustment of concrete capital goods in an endless repetition of the same process of investment and production. But this is only of importance in capital investments of relatively short duration.
>
> If, therefore, our analysis is only applicable within a fairly short period, then, strictly speaking, only short period capital goods (in other words, circulating capital) can be regarded as capital proper.

Indeed, Wicksell goes on to suggest that a Clark-like adjustment to a full long-run equilibrium is 'usually impossible': some forms of fixed capital can never be converted to 'circulating' uses; the 'volume' of fixed capital 'is, in most respects, on the same level as the unchanging original productive factors, labour and land'.[12]

The point comes up again later in Wicksell's (1934 [1919]) review of Cassel. Regarding the Austrian approach to distribution, in terms of the period of production, Wicksell (1934,I:237) notes:

> In the first place, this method gives us a purely theoretical insight into the very origin of interest; but practically, as I observed in my *Über Wert*, it has the serious drawback, which arises from the durability of certain capital investments, of presenting the process of successive re-adjustments, from which an equilibrium situation would ensue, as embracing an interval of time where centuries are mere episodes. This inconvenience, however, lies in the very nature of the subject-matter and cannot be avoided. For practical purposes we might of course confine our attention to shorter periods, and put particularly durable capital-goods in a group on their own as a kind of 'Rentengüter'—comparable to 'land' and the supply of natural forces. This procedure I there proposed and it is this which Professor Cassel now adopts, but of course we do not obtain more than a provisional equilibrium situation in this way.

There is, however, something perplexing in all this. In *VCR*, durable goods are excluded from the definition of capital to maintain the idea of an average period of production. That seems clear enough. In the *Lectures*, where the average period is dropped, the argument to exclude durables turns, instead, on the length of adjustment-time.[13] But the assumed context in this case is one of a stationary economy, where the adjustment to equilibrium is explicitly ignored (i.e., 'we do not take into account the actual transition stage'). Theoretically, Wicksell could have opened his system in the *Lectures*, with now no need to save the average period of production, to include the full adjustment of fixed capital goods, as in Clark or Walras. Uniform returns across *all* reproducible assets, would thus read as a condition of stationary equilibrium. If for no other reason than practicality, he thought it more appropriate to continue to work with a kind of 'short-period' or (again) 'provisional' long-run equilibrium throughout.[14]

The fourth and final component of the stationary state idea, for Wicksell (1934, I:202), is that net savings equal zero; 'the amount of capital must remain constant from year to year.' Capitalist consumption, once again, is given by rk in equation (4.1.13), leaving the stock of capital intact:

Capital is regarded in a stationary economy as capable of a certain, but on the whole not noticeable, oscillation, since it continuously increases by interest and is in the same way continuously decreased by the consumption of this interest. (1954:104)

And, a bit later, with regard to the level of subsistence for the 'entrepreneur':

The problem is *unsolved* so long as it is not known how much of his income the entrepreneur in question is accustomed to save. The *simplest* hypothesis is, however, that he does not save anything, but merely preserves his existing capital, that is to say, creates it afresh from period to period. But then his yearly subsistence and the average yearly return from his production (when he only works with his own means) are simply *identical magnitudes*. (1954:104)

The same arguments are carried over to the *Lectures*.[15] In a criticism of Walras, in particular, Wicksell (1934, I:171) makes the point formally explicit. With net savings as a function of the rate of interest, 'the assumption of *stationary* conditions' implies that S(r) = 0; expressing 'the truism that, in the stationary state, the inducement to new savings must have ceased.'

4.2b *The analytical core of* VCR[16]

With the conditions for a stationary equilibrium in mind, Wicksell proceeds to analyze the specific productivity conditions necessary to determine the rate of return on capital. In *VCR* the design and use of an average period of production is pre-eminent. The idea is to construct a system in which T becomes an 'independent concept' or a variable unrelated, savings behaviour aside, to the rate of interest.[17] Variations in T functionally determine r, the rate of return on the stock of capital. The latter, the size of the 'subsistence fund', 'advanced' to labour, consists of 'circulating' capital alone, *provisionally* given in amount.[18] The services of land and all durables, also treated as data, are assumed to be free:

If in the meantime we suppose that the services of the land and the use of the other rent-goods are free—the influence of these factors will be considered later—then Böhm-Bawerk in Volume III, Chapter V, of his book has taught us that with the help of the concept of the length of the production period a very simple relationship between the present position of wages and of capital-interest can be laid down, if the number of the available workers within an economy and the amount of capital are known. (1954:119)

A knowledge of $\bar{K}$, $\bar{N}$ and the function f(T) helps determine q, W and r. The value of T—the continuous-input/point-output case—is given by $\frac{1}{2}\tau$, indicating that the 'advance' to labour, on average, 'is consumed some time *before* the completion of the work (about half-way through the production)'.[19]

Wicksell (1954:121) beings at the firm-level, taking, successively, r and W as parametric; writing, with simple interest

$$Q = \tau WN(1 + \tfrac{1}{2}r\tau) \qquad (4.2.1)$$

for Q the 'value of the final product' and τ the period of production. It is not yet clear, at this point, whether Q is measured in terms of a *numéraire*, or, given a single produced good for the economy as a whole, in units of a homogeneous physical output.[20] Both sides of (4.2.1) are divided by N—implicitly assuming constant returns to scale—so that

$$q = W(1 + \tfrac{1}{2}r\tau) = W(1 + rT) \qquad (4.2.2)$$

is expressed in per capita amounts, showing 'product exhaustion'. Given r and setting $dW = 0$ or, alternatively, W with $dr = 0$, gives[21]

$$dq/d\tau = f'(\tau) = \tfrac{1}{2}Wr \qquad (4.2.3)$$

and, from (4.2.2),

$$W = q - \tau f'(\tau). \qquad (4.2.4)$$

In this section of *VCR*, Wicksell (1954:117) does not formally write a production function for net output; he does, however, informally state it:

> It is assumed in this case that within every single branch of business, the productivity of labour, for instance, the annual production of one worker, is, under otherwise constant circumstances, a *function* of the length of the production process—a function which increases with the length of this period but more slowly, so that the scale of the surplus returns becomes a decreasing one—an effect which entirely agrees with experience.

Note that the measure of q, once again, is in per capita amounts; essentially expressing equation (4.1.1) or, with the average period of production, $q = f(T)$.[22] The value of capital 'invested', on average, in each firm is given by

$$K = \tfrac{1}{2}WN\tau = WNT \tag{4.2.5}$$

so that (4.2.2), (4.2.3) and (4.2.5) are three equations—given K and N—in W, τ and r.[23] Wicksell (1954:126), solving (4.2.2) and (4.2.3) for W, and, using (4.2.5), writes the system, in reduced form, as

$$K = \frac{N}{2}(\tau q - \tau^2 f'(\tau)) \tag{4.2.6}$$

where 'q and $f'(\tau)$ must be thought of as known functions of τ'.

Aggregation is simply performed by assuming that the functions $f(\tau)$ are identical for each 'business':

> Capital and labour which are to-day employed in the manufacture of goods of a certain kind, can to-morrow have been partly transferred to other branches of business. Within the whole economy, however, the number of available workers and the total capital can be regarded as approximately given magnitudes. If, therefore, following Böhm-Bawerk's precedent, we may assume as a first approximation within all branches of the business *the same* productivity and the same *increase* in productivity when the length of the period of production is increased, then, obviously, our equations set forth above can be regarded as valid for the whole economy, since τ stands for the length of the period of production, q the yearly production of one worker, and W and r the wage and the level of interest. According to our assumptions, these values must be the same in all businesses. (1954:126–7)

The values $\bar{K}$ and $\bar{N}$, in other words, are fixed and given and—now clearly—a one-good assumption is adopted for the economy as a whole; q is the same for all firms. Although K as a stock is not yet explicitly written as physically identical to the flow of net output or q, it does read as a physical scalar. Equation (4.2.6), the reduced form, for Wicksell (1954:126), is one equation in one unknown: τ.

It is at this point, not surprisingly, in the move from the firm to the economy, that Wicksell (1954:127) recognizes what is now known as a 'Price-Wicksell Effect'. An exogenous increase in aggregate K 'is absorbed partly by the rise in wages, and partly by the lengthening of the period of production.' Although the rate of return on capital falls, the 'social' return $\hat{r}$, in equilibrium, is less than the 'private' rate of return, with wages held constant. Wicksell (1954:137–8) proceeds as follows. Differentiate equations (4.2.4) and (4.2.5), setting $N = \bar{N}$, to get

$$dK = \frac{N}{2}(Wd\tau + \tau dW) \quad (4.2.7)$$

and

$$dW = -\tau f''(\tau)d\tau \quad (4.2.8)$$

with $dW/d\tau > 0$.[24] An increase in K increases output by $Nf'(\tau)d\tau$, so that, using (4.2.7) and (4.2.8),

$$dQ/dK = 2f'd\tau/(Wd\tau + \tau dW) = 2f'(\tau)/(W - \tau^2 f''(\tau)) = \hat{r}; \quad (4.2.9)$$

noting that $r = 2f'/W$ from (4.2.3). Since $f''(\tau) < 0$, dQ/dK is always 'smaller than the rate of interest'; i.e., $r > \hat{r}$.[25] An equality between the 'marginal product of capital' and the interest rate applies to 'capital investments of the individual entreprenurs only'; equation (4.2.3) as an equilibrium statement for the economy as a whole is ambiguous.[26]

As a first extension to the above system [(4.2.1)–(4.2.5)], what he terms as 'a completion of Böhm-Bawerk's theory', Wicksell (1954:146–9) adds land as a second 'original' factor of production; still assuming that all durables are free goods.[27] Ground-rent is a return to the 'advance ' of land (paid out of capital), on average, for half the period of production. Simple aggregation is again assumed: 'landed property everywhere is of the same quality and in all branches of production an equally large area of land is required for each worker'. This time Wicksell formally writes a production function

$$q \equiv Q/N = f(\tau, l) \quad (4.2.10)$$

for $l = L/N$, the amount of land per capita; implicitly with constant returns to scale. Product-exhaustion gives

$$q = (W + Rl)\,(1 + \tfrac{1}{2}r\tau) \quad (4.2.11)$$

for R land-rent, with

$$dq/d\tau = \tfrac{1}{2}(W + Rl)r \quad (4.2.12)$$

and

$$dq/dl = R(1 + \tfrac{1}{2}r\tau) \quad (4.2.13)$$

or four equations in q, W, R, r, l and τ. Wicksell fixes the per capita amount of land and adds, comparable to (4.2.5), an equation

$$K = \tfrac{1}{2}N\tau(W + Rl) \qquad (4.2.14)$$

for 'the existing capital of the community', with $K = \bar{K}$. Setting $N = \bar{N}$ gives Q and L. 'The problem,' Wicksell concludes, 'is now solved in its entirety'; eight equations in Q, K, N, L, W, R, r and τ.

Finally, Wicksell (1954:153) sets out the conditions for the above systems [(4.2.1)–(4.2.5) and (4.2.10)–(4.2.14)] to work analytically. It is a striking passage in which the one-good assumption is formally adopted throughout:

> Let us first try to imagine what an economy must be like, if the equations [(4.2.10)–(4.2.14)] are to reflect the *true* play of economic phenomena. This requires, of course, that within the whole economy only one single consumption good, for instance corn, is produced. Wages, ground-rent and capital-interest are all received in the form of goods, that is to say, corn, and the capital itself consists of corn and the installments and tools necessary for the production of corn, which, however, we imagine as being so simple that they can be produced by the economies in question themselves and are of short duration. Durable goods are not produced at all. The economy must be a completely stationary one.

The fiction of 'installments and tools' is dropped several paragraphs later; so that q, W, R, r and, particularly, K are all measured in corn. Wicksell's reasoning is clear. The capital stock, as a physical scalar, is seen as independent of prices and incomes. It reads, *savings behaviour aside*, as a 'prior' accumulation of corn; given (in a stationary economy) in amount. Rates of return, ignoring land, for

$$\bar{k} = \bar{K}/\bar{N} = WT, \qquad (4.2.15)$$

are determined on marginal productivity grounds. Recall that T, with simple interest, is independent of r. Variations in T, or $f'(T)$, set the rate of return on capital, where the length of T depends on $\bar{K}$ and, relative to $\bar{N}$, the wage rate. In equilibrium, product-exhaustion [equation (4.2.2) with (4.2.3)–(4.2.4); or equation (4.1.22) with (4.1.24)–(4.1.25)], implies

$$q \equiv Q/\bar{N} = W + r\bar{k} \qquad (4.2.16)$$

setting W. A knowledge of $\bar{K}$, $\bar{N}$ (and $\bar{L}$) determines the relevant variables, all measured in corn.

4.2c VCR: *the 'missing equation'*

The foregoing defines the analytical core of *VCR*. There are two issues to consider: (a) identifying a 'missing equation' in the above systems; including whether or not Wicksell was aware that his equations must still be properly closed, and (b) whether or not it is appropriate to measure K as a physical scalar. In fact, the two are related. If Wicksell's systems are properly closed, the one-good assumption allows q, k, K, rK, and so on, to be represented as real aggregates; by, in effect, generating a surrogate function $q = f(k)$, for k measured as corn per capita.[28] Setting savings behaviour aside, k can be viewed, with simple interest, as independent of r. (Note Wicksell's reduced form, (4.2.6), as one equation in one unknown: τ.)

Undoubtedly, there is a 'missing equation' in the systems (4.2.1)–(4.2.5) and (4.2.10)–(4.2.14).[29] A period of production τ, implies that the capitalist, who 'advances' wage payments, must postpone consumption, essentially saving, on average, over T. In general, an intertemporal trade-off exists between the utility from consumption 'today' versus consumption 'tomorrow'. With a positive interest rate, the price of corn discounted to the present (i.e., in constant-value prices) will fall over time. Equilibrium, as in (4.1.7), or assuming two time-periods, requires that

$$\rho_t = \partial U/\partial c_t / \partial U/\partial c_{t+1} - 1 = r_t = (p_t - p_{t+1})/p_{t+1} \quad (4.2.17)$$

adding, for a given ρ_t, in discrete form, the needed extra equation; or $\rho = r$ in continuous terms. With current-value prices, or prices that are not discounted to the present, an $r_t > 0$, implies that the price of corn rises over time.[30] Variations in r, in turn, define a (gross) savings function **S**(r), with an equality between savings and investment properly closing the system. It follows that even with K as a physical scalar, the size of the 'subsistence fund' is a variable, depending on savings. An equation is needed to determine, endogenously, a constant stock of capital (as corn), a stock which is neither being currently consumed nor accumulated. With (4.2.17), or **S**(r) = WNT [see (4.2.20)], in other words, $\bar{K}$ in Wicksell's systems should read as an equilibrium amount K*, and not as exogenous—providing the rationale for the qualification, 'savings behaviour aside', throughout.[31] In Wicksell's reduced form, equation (4.2.6), T is independent of r, but K cannot be.

Was Wicksell aware of the issue or did he simply err? The evidence strongly indicates the former. As we have seen, Wicksell does indeed write his equations as if K is exogenous. But this does *not* imply that he failed to see the necessity of closing his systems—the block of equations (4.2.1)–(4.2.5) and (4.2.10)–(4.2.14)—with a savings function, or a condition defining intertemporal equilibrium, as in (4.2.17). The anomaly is explained by Wicksell's special handling of a stationary long-run equilibrium.

First, for analytical purposes, Wicksell separates, from an otherwise general equilibrium setting, arguments which determine the *size* of the capital stock from the determination, given a particular capital stock already accumulated, of the rate of interest. The point is made emphatically by Wicksell (1954:21–2):

> As I see it, these two questions—the question of the origin of capital interest and that of the origin of interest-bearing capital itself—however closely they are related, must in theory be first of all treated separately, just as, for instance, the theory of exchange is separated from the theory of production, though in reality exchange and production are almost always dependent on each other.

And, again, later in *VCR*:

> Another question which requires to be answered is why this stationary condition, or what comes to about the same thing *here*, a society in which there is only a slow progression, can be assumed to rule in theory as well as practice, and why the incomes of capitalists, landlords and workers are on the whole consumed instead of being hoarded and added to the stock of capital. And although this question is closely connected with the problem of interest, it remains nevertheless a question in itself. In my opinion, Böhm-Bawerk must be blamed for having mixed up the two questions of the origin of interest and the origin of interest-bearing capital itself—in his criticism of the older theories of interest as well as in his own positive presentation—instead of separating them in a truly scientific manner. (1954:113)

And, in the *Lectures*:

> Both logically and for the purposes of exposition it would seem right to begin by examining the effects of a given supply of capital already

accumulated, and *then* to inquire the causes which influence, and eventually alter, this supply. (1934, I:154–5)

In all of this, Wicksell is simply following Walrasian practice, i.e., building-up long-run equilibrium theory sequentially, with increasing generality. The theory of exchange—with given stocks of goods to be traded—is broadened to include a theory of production, where the number of goods, as produced output, can be varied. The theory of production, in turn, with given inputs, is broadened to accommodate a theory of accumulation, where the amount of capital, in particular, is variable. The assumption of a stationary economy allows Wicksell to keep the decision to save, which in general depends on r, distinct from what determines (i.e., the 'degree of roundaboutness') the rate of return to capital. *In short, Wicksell assumes that whatever the value of r, given by $f'(\tau)$, the amount of net savings, S(r), will be zero, leaving the stock of* K *intact.* Analytically, the 'feedback' from the rate of return to the amount of savings, and the amount of savings to the size of K, is, for the moment, ignored.

This perspective is especially prominent in systems using, with simple interest, the average period of production. Think of the issue in more modern neoclassical terms. Flow-equilibrium (with current-value prices) requires

$$S(r) = \dot{P}_i K_i + P_i[\dot{K}_i + \delta_i K_i] \qquad (4.2.18)$$

or that gross savings equals the change in the value of capital, for i different capital goods and δ_i the rate of depreciation.[32] A stationary equilibrium (or Wicksell's assumption of circulating capital goods alone), implies no capital gains/losses, $\dot{P}_i = 0$, and no net additions to the stock of capital goods, $\dot{K}_i = 0$, so that

$$S(r) = P_i \delta_i K_i. \qquad (4.2.19)$$

Finally, circulating capital, $\delta_i \equiv 1$, and the assumption of a single capital good (corn) gives, for Wicksell,

$$S(r) = K \left(= \tfrac{1}{2} WN\tau = WNT\right). \qquad (4.2.20)$$

The amount of savings *is* the stock of capital.[33] With simple interest, T, and hence K, on the production side, is independent of r; so that, as Wicksell sees it, $S \equiv K = WNT$. Each period rK is consumed, while K, generally an equilibrium stock, but written as if it is exogenous, is saved and 're-invested'. A stationary economy requires that *net* savings equal zero, regardless of r, leaving K unchanged.[34] The idea is simply to analyze the effect of a stock of savings-*cum*-capital, 'already accumulated'; to analyze 'the conditions necessary for the maintenance, or periodic renewal, of a *stationary state of economic relations*', savings behaviour aside.

The argument is enhanced by Wicksell's handling of Böhm-Bawerk's 'three grounds' for the existence of interest; a view which, combined with the above, accounts for the *appearance*, as Hirshleifer (1967:195) states it, of 'a one-sided productivity model' in *VCR* and the *Lectures*. Recall that stationary equilibrium, for Wicksell, implies, in terms of the 'first ground' for the existence of interest, that goods will exchange intertemporally at par; in equation (4.2.17), the rate of time preference, setting the 'marginal productivity of time' aside, becomes

$$\rho_t = \partial U/\partial c_t / \partial U/\partial c_{t+1} - 1 = (p_t - p_{t+1})/p_{t+1} = 0 \tag{4.2.21}$$

since $p_t = p_{t+1}$.[35] The value of ρ_t is *not* assumed, in and of itself, to be positive and constant. With income constant through time, as a condition of stationary equilibrium, there is nothing to explain a positive rate of time preference in terms of (4.2.21). For Wicksell (1934,I:155), this leaves only the 'third ground', as pre-eminent, to determine an interest rate; a positive r thus determines a positive ρ. Intertemporal preferences, the savings side of a more general setting, is, by design, subordinate. Although, in equilibrium, Wicksell affirms that $\rho = r > 0$, it is the productivity of capital that sets or 'leads' the rate of time preference.[36] Böhm-Bawerk's *agio*, or equation (4.2.17), matters especially—but not exclusively—in the context of additions to the stock of capital, but it is not 'a *main* ground of the phenomenon of interest' in a stationary economy.[37] The emphasis is necessarily on the 'degree of roundaboutness'.

But, again, this does not imply that Wicksell failed to see the necessity of closing his systems properly by determining the size of K endogenously as part of a general equilibrium problem. In *VCR*, the recognition first comes in the context of discussing the 'wages-fund doctrine', whereby per-capita wages are the result of dividing the capital destined for wage payments by the number of workers. 'Now it was pointed out,' Wicksell (1954:145) remarks, 'with good reason by the opponents of this theory, that the first of these magnitudes,' the amount of capital, 'is from the very start completely undetermined.' Indeed, the 'objection applies equally well to Böhm-Bawerk's theory'; the period of production idea as developed in *VCR*, where K reads as exogenous. It 'can only be removed', Wicksell argues, 'by a comprehensive theory of savings and capital formation'; by a device, a theory, which explains the size of K. The *Lectures* repeats the argument more pointedly. Wicksell (1934,I:202–03), in referring to a certain 'indeterminateness' in his systems, states:[38]

> This indeterminateness—which was inherent in our first main example, and even in the pure theory of production—is, of course, primarily due to the fact that capital, unlike labour and land, is not an *original* factor of production which can exist (even hypothetically) independently of, or antecedently to, production. Its origin and maintenance inevitably presuppose that production is taking place. But it also has another, more deep-seated, cause. In reality, the amount of capital is not determined by physical conditions, but by the equilibrium between psychical forces which, on the one hand, drive us to save and accumulate capital and, on the other, to consume already existing capital. In other words, the *accumulation of capital* is itself, even under stationary conditions, a necessary element in the problem of production and exchange.

And later:

> So far, our discussion has been based on the assumption that productive capital, like the other two factors, is constant. In reality, however, capital is not, like land—and for shorter periods, labour—physically limited. It can be increased at any moment by saving; it can be reduced by unproductive consumption. Neither is the supply of capital renewed in the same way as the supply of labour, by the work of nature—although it is natural to accumulate capital at certain periods of life (particularly middle age) and to consume it at others (early youth and old age). A rational theory of saving is thus necessary before we can clearly understand the conditions of a stationary society, with a constant supply of capital; and still more, of course, before we can understand and foresee the gradual changes in the amount of social capital. (1934,I:207)

This could not be clearer. Even with net savings assumed to be zero, a savings function—defining savings behaviour or the 'psychical forces' contained in intertemporal preferences, as in equation (4.2.17)—is required to close the system for stationary long-run equilibrium. The stock of capital does not 'exist (even hypothetically)' prior to production. The idea, along Walrasian lines, is to extend the problem to incorporate savings: 'we have now reached a point in our exposition at which this new factor forces itself upon our attention'; to explain the amount of K.[39]

There is further evidence. Although the emphasis, as discussed above, is on the 'third ground' throughout, Wicksell all but formally states (4.2.17) as an equilibrium condition. First, he acknowledges that a stationary economy gives constant relative prices over time, measured by the rate of interest:[40]

Under stationary conditions the exchange value of goods and services necessarily remains unchanged year after year, so that a person who, in one year, purchases labour and land in order to convert them into capital, intended for production in the following year, can always count upon obtaining more product, or value, than he himself paid out. This surplus is called interest. (1934,I:153)

To see this, assume two produced goods with prices p_i and p_j. The own-rate of interest for good i, in constant-value prices, is

$$r_{it} = (p_t^i - p_{t+1}^i)/p_{t+1}^i \tag{4.2.22}$$

in discrete form. Uniform returns on capital, used to produce both goods, imply

$$R_t = (1 + r_t) = p_t^i/p_{t+1}^i = p_t^j/p_{t+1}^j \tag{4.2.23}$$

for $r_t = r_{it} = r_{jt}$, leaving

$$p_t^i/p_t^j = p_{t+1}^i/p_{t+1}^j \tag{4.2.24}$$

or relative prices—as a condition of stationary equilibrium—unchanged over time. For $r_t > 0$, the undiscounted, or current-value price, for both goods i and j, increases over time, i.e., $P_{t+1}^i > P_t^i$, obtaining, *per*-Wicksell, 'more product,' or 'value' per-unit, than 'paid out' at t. The pattern of consumer demands, as reflected in relative prices over t, must conform to the rate of return on capital. This alone establishes a link between the preference and productivity sides of the model.

Second, Böhm-Bawerk's *agio*—the formula which 'governs the problem of interest in its whole extent'—implies a connection between the ratio of marginal utilities of present to future consumption, equated to relative prices, discounted to the present, i.e., 'according to contemporary valuation', in terms of r_t:[41]

Interest is an agio which comes into being when present and future goods are exchanged. It rests solely on the *relationship between present and future in human economy* and simply expresses the fact that *present goods* (at least according to the contemporary valuation) *are as a rule more valuable than future goods of the same kind and number.* (1954:107)

And:

The interest on the loan, just like the exchange value in the case of ordinary exchange, will depend on *two proportions of marginal utility*; that is to say, it will depend first on the proportion between the marginal utility of present and that of future goods *for the creditor*, and secondly on the proportion between the analogous marginal utilities *for the debtor*. Usually in this case the marginal utility of present goods for *both* will prove to be higher than the marginal utility of future goods of the *same* kind and *number*; so that interest almost always turns out to be positive—that is to say, it will be paid by the debtor. The proportion of marginal utility can finally become *identical* on both sides, but not the *proportion of the total utility*. This on the contrary, must always be different, if a loan is to take place at all, and in such a way that the debtor as opposed to the creditor always values present goods relatively higher. The interest which must really be paid will then fall somewhere or other between these two different valuations. (1954:107–8)

An equilibrium, in other words, gives

$$\rho_t = \partial U/\partial c_t / \partial U/\partial c_{t+1} - 1 = r_t = (p_t - p_{t+1})/p_t \qquad (4.2.25)$$

just as in equation (4.2.17). In constant-value prices, *per-*Wicksell, $r_t > 0$, implies that $p_t > p_{t+1}$. The exchange value of 'present goods' is greater than that of 'future goods of the *same* kind and *number*.'

Finally, in terms of savings behaviour, Wicksell (1934,I:208) ties equation (4.2.25) to the decision to forego present consumption:

Among the many influences affecting the accumulation of capital, the rate of interest is undoubtedly one . . . Theoretically, the individual should always carry his accumulation of capital (or it may be his consumption of capital) to the point at which present and future marginal utilities of the goods saved is equal.

The relation $\partial U/\partial c_t = (1 + r_t)\partial U/\partial c_{t+1}$ gives, in continuous form, with variations in r, a function $\mathbf{S} = \mathbf{S}(r)$. If net savings happen to be zero, the economy is stationary.

The fact that the *agio* mechanism is not more formally prominent in the structure of equations and unknowns, e.g., (4.2.1)–(4.2.5) and (4.2.10)–(4.2.14), in *VCR* and the *Lectures*, is undoubtedly explained by the intended preoccupation with the 'marginal productivity of time', or the 'degree of roundaboutness', a preoccupation stemming from a particular understanding of the importance of beginning with a stationary equilibrium.

At one point, in fact, Wicksell (1934,I:169) suggests that Böhm-Bawerk's 'first ground' can itself explain interest, irrespective of the return to capital:[42]

> [Böhm-Bawerk] here puts forward the doctrine that interest is originally an exchange phenomenon (and thus no longer exclusively the result of production and distribution)—it is the *agio* which arises in the exchange of present against future goods. This treatment may be justified, in so far as interest is undoubtedly a broader concept than productive capital itself. It can arise in a mere exchange of present against future goods or services *without* any intervening production and thus without any real accumulation or employment of capital.

But he does not place much weight on the argument, 'the proof is not quite convincing.' Recall that a stationary economy, for Wicksell, independent of r, implies that $\rho = 0$, with no tendency for 'savers' to predominate. To explain a rate of interest requires a market for capital and labour (i.e., a productivity criterion); starting with a given K and separating-out, provisionally, savings behaviour. This indeed is what Wicksell (1954:112) sees as one of his major clarifications of Böhm-Bawerk's work:

> Böhm-Bawerk himself, however, did not, or could not, remove the objection which he directed against this theory [the *agio*] in the first volume of his book—namely, that it explains at best the *physical*, but not the *value* production of capital. For the demand which he there makes of the productivity theorists was, after all, not to explain why present goods are higher in value than future goods of the same number and kind *according to the present valuation*—this (in so far as the above-mentioned fact [that interest will be positive] is generally true) Thünen's theory certainly explains as well as his own theory, though in a somewhat more concise manner—but *why the product of capital, when it becomes due, should be more valuable than the sacrificed capital commodity itself*. But Böhm-Bawerk has not explained this either; and it can after all only be explained *if one sets out from the assumption of a nearly stationary position of the economy*.

And, in the *Lectures*:

Böhm-Bawerk's real error—his cardinal error, as Bortkiewicz calls it—is that at this point in his exposition he seeks to solve the problem of the *existence* of interest—as distinct from its actual rate—without referring to the market for capital and labour. This error had already been pointed out by Walras and is, indeed, the only one of major importance which can be attributed to Böhm-Bawerk. (1934,I:171)

Wicksell's (1934,I:146) central concern—the *agio* aside—is, in short, to explain why the 'possession of capital' procures 'something more, namely a permanent income in the form of interest, either without sacrifice of capital or while capital is constantly being replaced.' The analytical procedure involves forcing, from an otherwise general equilibrium setting, a distinction between savings and the amount of K; a distinction between 'the origin of capital-interest' and 'the origin of interest-bearing capital itself'.

The point is strikingly seen in a criticism of Walras. Here, Wicksell (1934,I:171) notes that while S(r) = 0 (for S net savings, over and above, for example, **S**(r) = $\bar{K}$ = WNT), *is a condition of stationary equilibrium*, functionally S(r) alone cannot explain r:

Walras' formula for interest, as may be easily seen (cf. the preface to the second and subsequent editions of his *Éléments d'économie politique pure*) reduces itself, on the assumption of *stationary* conditions, simply to the equation S(r) = 0, in which S(r) is the amount of annual savings conceived as a function of the rate of interest r. In other words, it expresses the truism that, in the stationary state, the inducement to new savings must have ceased; but it affords no answer to the question why a given amount of existing social capital gives rise to a certain rate of interest, neither higher or lower. The importance of the time-element in production was never properly appreciated by Walras and his school. The idea of a *period* of production or of capital-investment does not, as we have said, exist in the Walras-Pareto theory.

That savings is recognized as a function of the rate of interest must imply that the stock of capital is endogenous to r. That Wicksell chose to concentrate on $f'(\tau)$ to determine the rate of interest, for a given K—'a question in itself'—and, hence, to ignore the 'feedback' from r to S, and S to K, can only have been a matter of focus or procedure, not principle. This procedure required, and was recognized by Wicksell to require, an extension to incorporate a savings function, or intertemporal preferences, to fully close the general model by determining the size of K.

Now, if Wicksell's systems are properly closed, Cass (1973) has shown, with a one-good assumption, that all the relevant variables—q, k, rK, and so on—appear as *real* aggregates; K reads as a physical scalar. It is required only that a given steady state (with exogenous growth rate g), be 'efficient', and subject to a neoclassical 'aging condition', i.e., $f'(\tau) > 0$ and $f''(\tau) < 0$, as indeed Wicksell has it. 'Efficiency' implies that intertemporal constant-value prices satisfy

$$r_t = (p_t - p_{t+1})/p_{t+1} \qquad (4.2.26)$$

just as in equation (4.2.17). Cass (1973:76) himself closes his system by maximizing the value of consumption each period

$$\max \sum_{t=0}^{\infty} p_t C_t \qquad (4.2.27)$$

subject to feasibility. But, in fact, any number of different conditions suffice, including the standard relationship, in continuous-time, between the rates of time preference, growth and the return to capital

$$\rho = r - g \qquad (4.2.28)$$

or simply, suppressing the preference side, r = g, or the 'Golden Rule' result for one-sector neoclassical growth, for a given g. Wicksell (1954:165) was aware that his systems could readily be extended to cover the case of continuous growth:

> If in all these relationships a certain *rate* of progression may be assumed to be given, then it is clear that equations of production and exchange can be laid down. We have, then, so to speak, a problem of dynamic equilibrium instead of a problem of static equilibrium with which to deal.

But there is nothing explicit in the account—here in the closing pages of *VCR*—to suggest that he saw the device of setting $r = \bar{g}$, adding the needed equation to (4.2.10)–(4.2.14), as a way of obtaining closure.[43] With land recognized as the ultimate fixed input (i.e., with a zero growth rate), the argument, in the context of the 'short-period', long-run equilibrium that Wicksell constructed, turns instead on the condition that $\rho = r$; with, as discussed above, r 'leading' the rate of time preference.[44]

4.2d *The two-consumption good, two-capital good model*

As a second extension to the system (4.2.1)–(4.2.5), now with land included [(4.2.10)–(4.2.14)], Wicksell adds an additional produced output, linen—also appearing as capital 'advanced' to labour—and assumes that the technology, the 'productivities', of producing corn and linen vary. The exercise is particularly notable since, with the stock of capital now defined across two 'circulating' goods, K no longer reads as a physical scalar. The aggregate value of capital is not only endogenous to savings behaviour, but to the relative price of corn in terms of linen. There is no confusion here.[45] Wicksell is clear that the value of K cannot, in general, be treated *a priori*, as a datum of the analysis.

Wicksell (1954:153–4) begins in the context of international trade. Each sector is independent—corn to produce corn, with labour and land, and linen to produce linen—with no transfer of resources between each economy. The idea, first, is to create a *direct* analog to (4.2.10)–(4.2.14), with K_1 and K_2, the respective amounts of corn-capital and linen-capital, as physical scalars:

> For each of these two economies there would then exist a system of equations similar to system (4.2.10)–(4.2.14). The constants of the equations—the number of workers, the area of land and the capital—as well as the form of the function of productivity f(l, t) are, however, different for both economies. Let the above-mentioned magnitudes be N_1, L_1, K_1 and q_1 for one economy and N_2, L_2, K_2 and q_2 for the other. When these magnitudes are inserted in equations (4.2.10)–(4.2.14) instead of N, L, K and q, we obtain from each of these equilibrium systems, by elimination of the remaining unknowns, first the length of the period of production t in question, then the values of the magnitudes W, R and r which we require to know. If these values are t_1, W_1, R_1 and r_1 for the first economy and t_2, W_2, R_2 and r_2 for the second, then $W_1N_1 + R_1L_1 + r_1K_1$ and $W_2N_2 + R_2L_2 + r_2K_2$ respectively express the quantities of goods which are produced every year in the two economies. Since, furthermore, the distribution of capital property and landed property within each economy must be assumed to be constant, we know how much of this production falls to each person's share.

The analogy is complete. The absence of mobility of resources between sectors/economies implies that all input rentals are industry-specific, with no uniform rate of return on capital. The addition of a second consumption good, however, implies an additional unknown: $\pi = p_1/p_2$, the relative price of corn to linen. To close, Wicksell simply adds, informally at this point, an equation expressing an equality between the supply and demand for one good; the same, for the second, is redundant.[46]

Next, Wicksell (1954:155) imagines 'that both economies are united *in a single economy*' with complete factor mobility, generating the two-sector, two-capital good case: corn, linen, labour and land to produce both corn and linen; allowing for a uniform rate of return on capital to be determined. In contrast to the initial situation (international trade), input coefficients and the value of aggregate capital are now endogenous:

If we wish to make use of two equilibrium systems [those from international trade] here, the difficulty arises that the magnitudes N_1, N_2, L_1, L_2, K_1, K_2 can no longer be assumed to be known; to begin with, we only know the sums $N_1 + N_2 = N$; $L_1 + L_2 = L$. As for capitals K_1 and K_2, neither they themselves nor their sum are known, strictly speaking. The national capital, in so far as it is free, consists here of *two* commodities, and its value can therefore only be determined after having found out their prices—that is to say, can only be expressed in one of these or in some other conventional medium of exchange.

The value of capital, in other words, is given by

$$K = \pi K_1 + K_2 \tag{4.2.29}$$

requiring a knowledge, for its determination, of both π and savings behaviour.[47]

Wicksell (1954:156–60) proceeds, *provisionally*, by fixing π and K. Factor-price equalization gives, in value terms, $W_2 = \pi W_1$, a uniform r, and so on. The remaining structure is a straightforward generalization of (4.2.10)–(4.2.14). The technology is represented by

$$q_i = f_i(l_i, \tau_i) \tag{4.2.30}$$

with product exhaustion giving

$$q_i = (W + Rl_i)[1 + \tfrac{1}{2}r\tau_i] \tag{4.2.31}$$

for $i = 1, 2$. First-order conditions for l_i and τ_i add four equations, comparable to (4.2.12) and (4.2.13). The value of capital in each sector becomes

$$K_i = \tfrac{1}{2}[N_i\tau_i(W + Rl_i)] \tag{4.2.32}$$

with resource constraints

$$\bar{N} = N_1 + N_2 \qquad \bar{L} = L_1 + L_2 \tag{4.2.33}$$

and (4.2.29) providing thirteen equations, in total, in q_i, N_i, L_i, K_i, τ_i, W, R, r, π and K. Wicksell, again, adds an equation—this time formally, in terms of utility functions defined over the consumption of corn and linen—expressing market equilibrium in the demand and supply for one of the goods.[48] All that remains—as throughout all of Wicksell's formal analysis—is to extend the system to incorporate intertemporal preferences, thereby tying-down the value of K with savings behaviour.[49]

4.3 The *Lectures*

4.3a *The transition from* VCR *to the* Lectures

There is a large measure of similarity in the capital theory contained in *VCR* and the *Lectures*, even with respect to the ordering of topics. All that is implied about stationary equilibrium, i.e., the separation of savings behaviour from capital 'already accumulated', constant income over time, net savings equal to zero, the concentration on the 'third ground' for interest, and so on, is carried over from one work to the other. There are, however, some important differences.

First, the use of compound interest throughout the *Lectures* means that T, the average period of production, is no longer independent of r; a point noted, but basically ignored, with the use of simple interest, in *VCR*. The measure of K in (4.2.5), as in equation (4.1.11), must depend on the return to capital. But, for Wicksell, something deeper is involved. Although the distinction between durable and 'circulating' capital goods, drawn from *VCR*, is retained in the *Lectures*, the importance of the presence of more than one 'circulating' capital good, advanced with possibly varying periods of production τ (what Wicksell now terms the 'period of investment'), is thoroughly recognized. As a result, net output can no longer be expressed as a physical scalar or as a simple function of T, aggregated over firms, since the one-good assumption is dropped and production periods, in general, do not uniformly overlap.

To avoid these problems—though not the perceived necessity of measuring q and K strictly in value terms—Wicksell, in much of the formal analysis (such as the 'wine models'), adopts the point-input/point-output case. A uniform distribution of production processes is, for the most part, still assumed but labour, with land, is not uniformly applied over t. By construction, point-input implies that the average period of production is always identically equal to the period of investment—T is theoretically superfluous—while point-output excludes the production of durables or fixed capital. The focus largely, although not exclusively, is on the level of the firm: (a) the exchange value between any two 'circulating' capital goods—or with one 'circulating' capital good (e.g., wine), the exchange value across successive vintages—as produced output, by the firm, is taken as given; (b) either τ, for simplicity, is assumed to be identical across firms, or different investment periods, specific to each firm, cannot be aggregated and represented by a function $f(\tau)$; and (c) the 'Price Wicksell Effect'—much more prominent in the informal discussion contained in the *Lectures* than in *VCR*—implies that the theory regarding the 'marginal product of capital' only rigorously holds 'from the point of view of the entrepreneur, to whom wages and rent are data, determined by the market.'[50]

This leads to a second and crucial point. The main conceptual difference between *VCR* and the *Lectures* is Wicksell's (1934, I: 165) desire, in the *Lectures*, to represent 'social' capital—a 'complex' of quantities or a stock of heterogeneous non-durables—as (by necessity) an aggregate-value. In fact, the only point in the *Lectures* where Wicksell keeps the one-good assumption, with respect to K, is in the 'wine models', where 'circulating' capital consists of wine alone. But even here, he proceeds, right from the outset, strictly in value terms: K, with r continuously compounded, is not represented as a stock of wine, but as a sum of value. Wages, although paid or 'advanced' in wine, are exchanged for a basket of goods at internationally given prices. The idea, in short, is to measure K not as a physical scalar but only in terms of equilibrium prices, as a 'single sum of value, a certain amount of the medium of exchange, money.'[51] As Wicksell (1934, I:149) explains:

Whereas labour and land are measured each in terms of its own *technical* unit (e.g., working days or months, acre per annum) capital, on the other hand, as we have already shown, is reckoned, in common parlance, as a sum of *exchange value*—whether in money or as an average of products. In other words, each particular capital-good is measured by a unit extraneous to itself. However good the practical reasons for this may be, it is a theoretical anomaly which disturbs the correspondence which would otherwise exist between all the factors of production. . . . If capital also were to be measured in technical units, the defect would be remedied and the correspondence would be complete. But, in that case, productive capital would have to be distributed into as many categories as there are kinds of tools, machinery, and materials, etc., and a unified role of the treatment of capital in production would be impossible. Even then we should only know the *yield* of the various objects at a particular moment, but nothing at all about the value of the goods themselves, which it is necessary to know in order to calculate the rate of interest, which in equilibrium is the same on all capital.

The perspective is essentially Walrasian, in particular Part IV of the *Elements*. Each fixed capital good, given in arbitrary amounts, and ignoring depreciation—just as in Wicksell's provisional or 'short-period' long-run equilibrium—earns a specific rental, or yield. As such, there is nothing, structurally, to explain a uniform r or prices for durable goods.[52] The procedure that Wicksell adopts, as a result, is to consider various 'circulating' capital goods—or the value of 'wine', as a single 'circulating' capital good, defined across vintages—as separate from durables, with input-specific returns; as a sum of value K, transferable between uses, earning a uniform rate of return, or rK. A value for r, in turn, gives imputed prices, on the capitalized value of rentals, for each fixed or durable capital good.

This explains a great deal. The analytical core of *VCR*—a single, 'circulating' capital good or corn, which generates a stationary flow of physically identical net output, with simple interest accruing—is all but gone from the *Lectures*. Capital, rather than

$$K = \tfrac{1}{2}\tau(WN + RL) = T \cdot (WN + RL) \qquad (4.3.1)$$

is now measured by

$$K = (WN + RL) \int_0^{\tau} e^{rt} dt \qquad (4.3.2)$$

or the value of time-augmented wage and rental payments, with *r* as a market rate of interest, continuously compounded, equal to the rate of return on capital, for an overall—though (possibly) variable across different production processes—period of investment τ.[53] At the firm level, with a uniform τ, the function $f(\tau)$ replaces $f(T)$ throughout. All 'circulating' capital goods, indeed all produced capital goods, 'however different they may appear, can always be ultimately resolved' into payments to labour and land that 'belong to *earlier years*'.[54] The idea of K as a 'subsistence fund', since it no longer reads as a physical scalar, is dropped in favour of dated labour- and land-inputs.[55] For Wicksell (1934, I:246), capitalist production, rather than the idea of capital itself, becomes pre-eminent; at the aggregate level, only the marginal productivities of labour, land and time matter.[56]

4.3b *The 'wine models'*

In the first of Wicksell's (1934, I:172–81) two 'wine models'—the production of grape-juice, stored-to-age in successive vintages—land is considered to be a free good. The total supply of labour, given exogenously, is '*used only once*, as it were in an indivisible moment of time', to harvest grapes and provide juice, at $t = 0$; the point, in this case, furthest away from matured wine as realized output. Stored wine is 'spontaneously matured by *free* natural forces' over time, indexed by t, for τ periods, the length of the 'period of investment'. Wicksell assumes a closed economy, which 'produces only a single commodity—let us say a certain kind of wine—in exchange for which it obtains all other commodities from neighboring countries'. Production processes, as successive vintages, are uniformly distributed over time, each having the same 'period of investment'. The stock of 'circulating' capital, as stored wine, is used to 'advance' wages over τ. Its exchange value—Wicksell is clear on this point—is endogenous to the market prices for wine of each vintage.[57]

Let P_0 designate the undiscounted value of grape-juice at $t = 0$ and P_t the undiscounted value of matured wine at each t. Wicksell (1934, I:172) assumes that P_t is *given* for all t, 'determined in advance on the market', in such a way that it '*increases* continuously with the *age* of the wine.' This, indeed, is an important step, since current-value prices give

$$r_t = (P_{t+1} - P_t)/P_t \tag{4.3.3}$$

as the market rate of return—what Wicksell terms the market 'real rate of interest'—in effect determining r_τ at τ, once τ is known.[58] In equilibrium, for Wicksell (1934, I:178, 180–81), r_t must also equal the rate of return on the stock of capital. The value of capital, in turn, given the 'period of investment', is simply the discounted value of wine

$$K_t = P_\tau \cdot 1/(1 + r_\tau)^{\tau - t} \tag{4.3.4}$$

summed over all vintages, e.g., for $\tau = 4$, $K_3 = P_4 \cdot (1 + r_4)^{-1}$, and so on. Thus the selling price of three-year wine (its capital-value) equals the selling price of four-year wine, discounted at $(1 + r_\tau)$, for

$$P_\tau/P_{\tau-1} \geqslant (1 + r_\tau) \geqslant P_{\tau+1}/P_\tau \tag{4.3.5}$$

with time, as in Wicksell (1934, I:174–5, 177), measured in 'years'. The 'period of investment', in short, as an unknown, is '*the only variable dimension of capital*'.

Formally, represent the value-production of wine by

$$P_\tau Q = F(N, \tau) = N \cdot f(\tau) \tag{4.3.6}$$

for Q the amount of grape-juice provided by labour at $t = 0$ and $P_\tau Q$ the value of grape-juice-*cum*-wine at τ. Wicksell (1934, I:178), at this point, suppresses both Q and N, assuming that one unit of labour produces one unit of grape juice, and writes equation (4.3.6) as

$$P_\tau = f(\tau) \tag{4.3.7}$$

so that the analytical focus turns on the per-unit value of wine. The relationship between the value of grape-juice and P_τ

$$P_\tau = f(\tau) = P_0 \cdot (1 + r_\tau)^\tau \tag{4.3.8}$$

depends on the length of the 'period of investment' and the resulting rate of interest; or

$$P_\tau = P_0 e^{r\tau} \tag{4.3.9}$$

in continuous form. The value of P_0, again, as in equation (4.3.5), is simply P_τ, discounted by $(1 + r_\tau)$, for τ periods.

The 'marginal product of time'—the rate of return r_τ on the stock of capital—is the extra value obtained by 'prolonging the period of storage (i.e., the period of production or capital investment)'. Variations around τ, using (4.3.8), give

$$f(\tau)/f(\tau - 1) \geqslant (1 + r_\tau) \geqslant f(\tau + 1)/f(\tau) \tag{4.3.10}$$

or, using (4.3.9),

$$r = f'(\tau)/f(\tau) \tag{4.3.11}$$

for $f'(\tau) = \partial f(\tau)/\partial\tau$. The 'social value of capital', summed over all vintages, finally, is

$$K = P_0 \sum_{t=0}^{h} (1 + r_t)^t \tag{4.3.12}$$

for $h = \tau - 1$, the number of analytic time-periods; or

$$K = P_0 \int_0^\tau e^{rt}dt = (P_\tau - P_0)/r \tag{4.3.13}$$

in continuous form. Note, as well, that the expression for the 'marginal product of capital' designates a 'Price-Wicksell Effect'

$$dP_\tau/dK = r + [K - P_0\tau](dr/dK) \tag{4.3.14}$$

since $K > P_0\tau$, or $dP_\tau/dK < r$. Most importantly, for Wicksell, the system (4.3.7), (4.3.8) [or (4.3.9)], (4.3.10) [or (4.3.11)] and (4.3.12) [or (4.3.13)] is closed; representing four equations in τ, P_0, K and r_τ. Adding equation (4.3.5), with a knowledge of τ, gives the relative price $P_{\tau+1}/P_\tau$. The rate of return r_τ must conform to given market prices across vintages of wine. This is a crucial point, since the usual interpretation of Wicksell considers the wine model to be open, with Wicksell closing by taking the value of capital as exogenous. But, in fact, this is not the case here; it is not Wicksell's procedure. The value of K is not taken as a datum of the analysis. It is endogenous to τ and the resulting value of r_τ.

All of this can be placed in a more standard setting, using (4.3.6), in terms of the amount of labour and wage payments.[59] In place of (4.3.9) write

$$P_\tau q = f(\tau) = We^{r\tau} \tag{4.3.15}$$

for $q = Q/N$. Indeed, Wicksell (1934, I:172–3) remarks that P_0 can be 'entirely resolved' into total wage payments WN. Equation (4.3.11) remains. The value of capital is

$$K = WN \int_0^\tau e^{rt}dt = WN(e^{rt} - 1)/r \tag{4.3.16}$$

where, combining (4.3.11) and (4.3.15)–(4.3.16), gives

$$P_\tau Q = WN + rK \tag{4.3.17}$$

or 'product exhaustion' in value terms. In discrete form, ignoring the inequality, equation (4.3.10) becomes

$$r_\tau = [f(\tau + 1) - f(\tau)]/f(\tau) = (P_{\tau+1} - P_\tau)/P_\tau \tag{4.3.18}$$

or $f'(\tau)/f(\tau) = \dot{P}(\tau)/P(\tau) = r$, for the market rate of interest equal to the rate of return on capital. Finally, fix the amount of labour

$$N = \bar{N} \tag{4.3.19}$$

exogenously. Equations (4.3.6), (4.3.11), (4.3.15), (4.3.16), (4.3.18) and (4.3.19) are six in number to determine Q, N, K, W, τ and r.[60] With τ, the value of $(1 + r_\tau)$ must conform to the given value $P_{\tau+1}/P_\tau$.

The only step remaining, for Wicksell, is to tie $P_{\tau+1}/P_\tau$ to intertemporal preferences, defined by a utility function over the consumption of wine, or $U = U(c_\tau, c_{\tau+1})$. In constant-value prices, the condition, or the 'formula', which 'governs the problem of interest in its whole extent':

$$\rho_\tau = \partial U/\partial c_\tau / \partial U/\partial c_{\tau+1} - 1 = (p_\tau - p_{\tau+1})/p_{\tau+1} = r_\tau \tag{4.3.20}$$

allows for prices to be determined endogenously. Although (4.3.20), as discussed above, is standard teaching in both *VCR* and the *Lectures*, at this point, in the above systems, Wicksell simply prefers to take market prices—in current-value terms—as exogenous to close. The 'third ground' for the existence of interest in terms of τ, or $r = f'(\tau)/f(\tau)$, still predominates.[61] Note, however, that equation (4.3.5) or (4.3.18) is still required, in equation (4.3.12), to find the discounted value of the stock of capital, as an endogenous variable.

In Wicksell's (1934, I:181–2) second 'wine model', land, also solely employed at $t = 0$, is incorporated, with K measured as time-augmented wage and rental payments.[62] Wicksell explicitly assumes constant returns to scale here and, as a result, notes (as a standard outcome) the loss of an independent equation. There is, however, an additional complication. The 'periods of production' of labour- and land-*cum*-wine, τ and θ respectively, are initially assumed to vary. With L and R, an extra unknown, or θ, is added to Wicksell's first 'wine model'.

Represent the value-production of wine by

$$P_{\tau/\theta}Q = F(N, L, \tau, \theta) \qquad (4.3.21)$$

where the chosen subscript on P depends on which period of production has the greatest length. 'Product exhaustion', comparable to equation (4.3.15), gives

$$F(N, L, \tau, \theta) = WNe^{r\tau} + RLe^{r\theta} \qquad (4.3.22)$$

for r the rate of return on capital, with the total amounts of labour and land fixed

$$N = \bar{N} \qquad L = \bar{L} \qquad (4.3.23)$$

exogenously. First-order conditions for direct and time-augmented labour and land inputs are

$$F_N = We^{r\tau} \quad F_L = Re^{r\theta} \quad F_\tau = rWNe^{r\tau} \quad F_\theta = rRLe^{r\theta} \qquad (4.3.24)$$

for $F_N = \partial F/\partial N$, $F_\tau = \partial F/\partial \tau$, and so on. The 'money value of social *capital*', in this case, is given by

$$K = WN \int_0^\tau e^{rt}dt + RL \int_0^\theta e^{rt}dt \qquad (4.3.25)$$

or the sum of τ-years' wages and θ-years' rent plus interest accruing at the rate r. Leaving $P_{\tau/\theta}$ aside, there are nine equations in Q, N, L, τ, θ, W, R, K and r.

Constant returns to scale, however, leaves (4.3.22) redundant to first-order conditions, losing an equation; as Wicksell (1934, I:182) remarks, (4.3.22) 'is an identity so long as $F(\cdot)$ is a homogeneous and linear function in N and L'. To capture this, rewrite the system as an analog to equations (4.3.15)–(4.3.19) assuming, for the moment, that the 'periods of investment' for labour and land are identical, or that $\tau = \theta$. The value of matured wine is

$$P_\tau q = f(l, \tau) \qquad (4.3.26)$$

for $q = Q/N$ and $l = L/N$; or

$$P_\tau q = f(l, \tau) = (W + Rl)e^{r\tau} \qquad (4.3.27)$$

with $f_l = Re^{r\tau}$ and $\partial f(\tau)/\partial\tau \,/\, f(\tau) = f'(\tau)/f(\tau) = r$ as the remaining first-order conditions. The value of capital

$$K = (WN + RL) \int_0^\tau e^{r\tau}dt = (WN + RL)\,[e^{r\tau} - 1]/r \qquad (4.3.28)$$

returns equation (4.3.2) and $P_\tau Q = WN + RL + rK$. Wicksell assumes, once again, that P_τ is a known function of τ or that market prices for matured wine are given, so that, in discrete form, equation (4.3.18), or $f'(\tau)/f(\tau) = \dot{P}(\tau)/P(\tau) = r$, remains. In this way an equality between the rate of interest and the rate of return on capital is established. With (4.3.23), there are eight independent equations in eight unknowns: Q, N, L, τ, K, W, R, r.[63] A knowledge of τ gives, in discrete terms, $(1 + r_\tau) = P_{\tau+1}/P_\tau$. In terms of the value of per-unit grape-juice, P_0 can be 'totally resolved' into payments (WN + RL), with $P_\tau = P_0 \cdot (1 + r_\tau)$, per unit of wine.[64]

It is clear, however, that if τ does not equal θ, the system remains open. In this case either an additional equation is required or—'with one degree of freedom'—one of the endogenous variables must be arbitrarily fixed. Wicksell (1934, I:182) proceeds, at this point, with the latter option, closing the system either, for example, 'by supposing τ or θ (or some particular relation between them) to be given, or else by some supposition as to the money value of social *capital*'. It should be stressed that this is the *only* point in the 'wine model' discussion where the device of treating K as a datum is considered; the result of adding θ as an unknown. The argument is clearly made in provisional terms. With $\tau \neq \theta$, the value of K—conditionally fixed—depends on an equation, comparable to (4.3.20), expressing savings behaviour or intertemporal preferences. Assuming, alternatively, that $\tau = \theta$ leaves the value of capital properly endogenous, given equilibrium prices, to τ or r_τ. The value of P_τ, however, must still be explained by the 'first ground' for the existence of interest—by savings behaviour and a condition specifying that $\rho = r$, in constant-value prices—a point repeatedly affirmed by Wicksell in *VCR* and the *Lectures*.

4.3c *The two-consumption good, two-capital good model*

Conceptually, the two-consumption good, two-capital good model, contained in the *Lectures*, is identical to the system found in *VCR*.[65] A single economy, with complete factor mobility, uses two 'circulating' capital goods—corn and linen—with labour and land, to produce both corn and linen as consumption goods. The aggregate value of capital, again, is not only endogenous to savings behaviour, but to the relative price of corn in terms of linen; or p_1/p_2. The main difference in the *Lectures* is that

Wicksell (1934, I:196–205) measures capital—as 'advanced' corn and linen—solely in terms of *dated* labour- and land-inputs. Output, as a result, is a function of '*current* and *saved up*', or direct and past, amounts of labour and land. The system moves backward in time. Define $t = 0$ as the moment when produced output is realized, for τ the length of the 'period of investment', and index past labour and land, for $t > 0$, by N_t and L_t. Direct labour N_0, for example, is applied at $t = 0$; labour from one period in the past is N_1, from two periods ago N_2, and so on; where the 'earliest' application of labour, in discrete time-periods, is given by $N_{\tau-1}$. Interest is continuously compounded. Prices are expressed in constant-value terms.

Index consumption goods by j, so that

$$Q_{0j} = F_j(N_{0j}, L_{0j}, N_{tj}, L_{tj}) \tag{4.3.29}$$

represents the output of corn and linen, for functions F_j homogeneous and linear. There are 4τ first-order conditions, of the form

$$\partial Q/\partial N_{0j} = W \quad \partial Q/\partial L_{0j} = R \quad \partial Q/\partial N_{tj} = W(1 + r)^t$$
$$\partial Q/\partial L_{tj} = R(1 + r)^t \tag{4.3.30}$$

for t defined from $t = 1$ to h, for $h = \tau - 1$. Wicksell (1934, I:204) fixes the total amounts of labour and land—with uniformly distributed stages of production operating simultaneously—exogenously, or

$$\bar{N} = N_{0j} + \sum_{t=1}^{h} N_{tj} \quad \bar{L} = L_{0j} + \sum_{t=1}^{h} L_{tj} \tag{4.3.31}$$

summed over j. The value of capital is given by

$$K = W \sum_{t=1}^{h} N_{tj}(1 + r)^t + R \sum_{t=1}^{h} L_{tj}(1 + r)^t \tag{4.3.32}$$

summed, again, over j. Finally, given 'personal preferences in consumption', Wicksell solves for the relative price p_1/p_2, by adding an equation expressing an equality between the supply and demand for corn; a condition of market equilibrium. Price equations, showing 'product exhaustion', are

$$p_j Q_{0j} = WN_{0j} + RL_{0j} + W \sum_{t=1}^{h} N_{tj}(1 + r)^t + R \sum_{t=1}^{h} L_{tj}(1 + r)^t \tag{4.3.33}$$

for each j, implying a uniform rate of return r on the value of capital.

With the condition for market equilibrium, (4.3.29)–(4.3.32) represent $4\tau + 6$ independent equations in $4\tau + 7$ unknowns: 4τ input coefficients $[N_{0j}, L_{0j}, N_{tj}, L_{tj}]$, Q_{0j}, W, R, K, p_1/p_2, and r.[66] The system moves 'with one degree of freedom', i.e., it is 'short' one equation for the number of unknowns.[67]

Wicksell (1934, I:204–5) proceeds—just as, at this point, in *VCR*—by arbitrarily fixing the value of capital. *But the provisional nature of such a solution is explicitly recognized here; in a more general setting, the value of K depends on both relative prices and savings behaviour.* Wicksell addresses the issue in terms of two different kinds of 'indeterminateness' contained in the system (4.3.29)–(4.3.32). First, as Wicksell (1934, I:202) remarks:[68]

> It may well be questioned what importance we are to attach to the claim that, under stationary conditions, the amount of capital must remain constant from year to year. But here we must distinguish two different things. In equilibrium, the capital employed in production has already assumed a certain technical dimension and composition, as well as a certain exchange value (expressed in terms of one of the commodities). It can now be asserted that, so long as capital of this magnitude and composition, or even of this exchange value, is maintained and utilized from year to year, equilibrium cannot be disturbed if, from the beginning, the other conditions of stability are fulfilled. But it would clearly be meaningless—if not altogether inconceivable—to maintain that the amount of capital is already fixed *before* equilibrium between production and consumption has been achieved. Whether expressed in terms of one or the other, a change in the relative exchange value of two commodities would give rise to a change in the value of capital, unless its component parts simultaneously underwent a more or less considerable change. But even if we conceive of capital genetically, as being a certain quantity of labour and land accumulated in different years, a change in the value of commodities would also alter the conditions of their production and thus necessitate a larger or smaller change in the composition of capital.

And, second:

> This indeterminateness—which was inherent in our first main example, and even in the pure theory of production—is, of course, primarily due to the fact that capital, unlike labour and land, is not an *original* factor of production which can exist (even hypothetically) independently of, or antecedently to, production. Its origin and maintenance inevitably presuppose that production is taking place. But it also has another, more

deep-seated cause. In reality, the amount of capital is not determined by physical conditions, but by the equilibrium between psychical forces which, on the one hand, drive us to save and accumulate capital and, on the other, to consume already existing capital. In other words, *the accumulation of capital* is itself, even under stationary conditions, a necessary element in the problem of production and exchange. (1934, I:202–3)

This could not be more clear. The system (4.3.29)–(4.3.32) is 'indeterminate', first, in the sense that the value of capital [measured as $K = \pi K_1 + K_2$, in the final stage of production, as in *VCR*, or in terms of dated labour- and land-payments, as in equation (4.3.32) or (4.3.33)] is endogenous to equilibrium prices; specifically to the exchange value $\pi = p_1/p_2$ of corn to linen. A change in π changes the value of K. The amount of capital cannot be 'fixed *before* equilibrium' has been 'achieved'; to do so 'would clearly be meaningless—if not altogether inconceivable'.

The system is also 'indeterminate' in a second sense. The value of capital is endogenous to savings behaviour or to the 'psychical forces' expressed by intertemporal preferences, as in equation (4.2.17), for the consumption of corn and linen over time. The amount of capital cannot 'exist (even hypothetically) independently of, or antecedently to, production.' A market equilibrium equating the supply and demand for corn, and an equation such as $\rho = r$ to determine the amount of savings turned into capital, are required to close (4.3.29)–(4.3.32).[69] Although Wicksell is not specific at this point in the *Lectures*, it is clear that savings

$$\mathbf{S} = \mathbf{S}(W, R, p^1_\tau, p^2_\tau, p^1_{\tau+1}, p^2_{\tau+1}, r_\tau) \qquad (4.3.34)$$

is considered to be a function of all incomes and the prices of corn and linen over time. The own-rate of interest for corn

$$r_{1\tau} = (p^1_\tau - p^1_{\tau+1})/p^1_{\tau+1} \qquad (4.3.35)$$

and, as Wicksell (1934, I:175) argues, arbitrage '*in transactions between holders*' of corn and linen, imply that

$$R_\tau = (1 + r_\tau) = p^1_\tau/p^1_{\tau+1} = p^2_\tau/p^2_{\tau+1} \qquad (4.3.36)$$

for $r_{1\tau} = r_{2\tau} = r_\tau$; expressing both uniform rates of return—the same expression, most importantly, used by Wicksell to tie the 'first ground' for interest to ρ_τ—and

$$p^1_\tau/p^2_\tau = p^1_{\tau+1}/p^2_{\tau+1} \qquad (4.3.37)$$

or constant relative prices over time, as a condition of stationary equilibrium. Savings, as a result, is simply a function of

$$\mathbf{S} = \mathbf{S}(W, R, \pi, R_\tau) \qquad (4.3.38)$$

since R_τ and π capture all relative prices.

The system of equations, that solves for market values and input returns must guarantee, as discussed above, a π (e.g., the value of wine across successive vintages) that conforms to $(1 + r_\tau)$. If net savings $S(r_\tau)$—often for Wicksell simply given as a function of the rate of interest—happen to be zero (the meaning of a stationary economy) the value of capital, *endogenous* to π and r_τ, remains intact, as 'an equilibrium that cannot be disturbed'. Only then can the distinction between the 'origin of capital-interest and that of interest-bearing capital itself' be drawn.[70]

4.4 Conclusion

Wicksell, except in *provisional* terms, with substantial qualification, does not take the value of aggregate capital as given to solve for a stationary equilibrium. Although the emphasis is necessarily on the productivity-side of the model, Wicksell, contrary to Hirshleifer, Garegnani, Petri, Howard, etc., sees the need and sets out the conditions required to close *all* of his systems of equations properly. The bottom line is straightforward: in his approach to the theory of long-run equilibrium, Wicksell is as thoroughly neoclassical as Hicks (1946) and Hirshleifer (1967). A savings function and intertemporal preferences matter in the determination of the value of capital.

Notes to Chapter 4

1 The principal sources for this section are Dorfman (1959a), Hirshleifer (1967), Kuenne (1971), Lutz (1968), Samuelson (1982b), Swan (1956) and von Weizsäcker (1971). Standard references include Dorfman (1959b), Gifford (1935), Kuenne (1962), Neuberger (1959) and Stigler (1961).

2 Note the correspondence to standard one-sector neoclassical theory. The function $q = f(k)$ replaces (4.1.21), with K in $k \equiv K/N$, an exogenous physical scalar; and $q = W + rk$, $r = f'(k)$, and $W = f(k) - kf'(k)$, two of which, given constant returns to scale, are independent, replace (4.1.22), (4.1.24) and (4.1.25).

3 See Wicksell (1954 [1893]:22, 103, 119, 120) and Wicksell (1934, I:7, 153, 193). Note the comparative-statics method, e.g., Wicksell (1934, I:152): 'We shall assume *stationary conditions* as the foundations of our observations. This will not prevent us from considering changes in the quantities concerned, provided that we do not take into account the actual transition stage, which is a much more complicated problem, but assume that these changes have already become final, so that "static equilibrium" (a stationary state) is again restored.'

4 See Wicksell (1954:21–2) and Wicksell (1934, I:155).

5 Wicksell (1954:107).

6 See Wicksell (1954:22, 120) and, from the *Lectures*, Wicksell (1934, I:7):'We shall treat the subject from the static point of view, i.e., we shall assume, in principle, a society which retains unchanged from year to year the same population, the same area of territory and the same amount of capital, and remains on the same level of technical achievement.'

7 The use of the term 'approximately', which appears several times in this connection, is apparently motivated—theoretical purpose aside—by the recognition that capital, as well as labour, is not, in actual fact, fixed in amount. See Wicksell (1934, I:207).

8 Indeed, this recognition leads Wicksell (1954:108–110) to deny Böhm-Bawerk's 'first main ground' for the existence of interest, i.e., 'the difference in the circumstances of want and provision at different periods of time.' As Wicksell notes: 'In a stationary economy (which in my opinion must always be considered first as the simplest case), needs and their satisfaction are to be understood as, on the average, constant magnitudes. In such an economy also, it is true, several persons, or whole age-groups, could expect a more abundant satisfaction in the future than now. But besides these there are other individuals for whom the opposite is true; so that it seems as if, under this assumption, supply of, and demand for, present goods against future goods must equal each other at *par*.' With a constant stream of net income, in other words, there is no necessity for net savers to predominate and, as a result, no explanation (were it not for a 'marginal productivity of time'), as Wicksell sees it, for a $r > 0$. In terms of a corn own-rate of return, $p_t = p_{t+1}$ for all t, or $r = 0$, so that present and future goods exchange, intertemporally, at *par*. For a stationary equilibrium, this leaves the 'third ground', i.e., 'that time-consuming, round-about methods of production are more productive', as the only determinant of a positive rate of interest. On constant relative prices over time, defined over a number of goods, see Section 4.2c below.

9 Note Wicksell (1954:105): 'It seems best to me for the purposes of the following investigation to class the different capitals simply *according to their durability*. In what follows I shall call the highly durable goods rent-goods, whether they are products themselves, or, like virgin soil, goods furnished by nature itself and whether they yield useful services spontaneously or only by the addition of human labour. Consumable or quickly exhausted production or consumption goods, so long as the latter are not yet in the hands of consumers, I shall call capital-goods or capital in the narrower sense.'

For the most part in *VCR*, the stock of 'circulating' capital is assumed to be homogeneous, as bushels of corn. The number of each type of durable capital good is kept constant by assuming that depreciation allowances are always 're-invested', keeping each physically intact. As Wicksell (1954:162) states it: 'In the stationary economy such goods are not produced at all, but are kept in the same good condition.'

10 See von Weizsäcker (1971:67–72).

11 Wicksell often misses the point that with durables, part of output must be used to cover depreciation or replace worn-out 'machines', i.e., output must include more than corn.

There is an additional case made in *VCR*. Wicksell (1954:118) notes that the more durable a capital good, the less 'the original cost of construction' affects current rentals; the less 'prospective returns in some distant future' affect its 'present capital-value or profitableness'. As a result, Wicksell (1954:119, 163) concludes, without much explanation, that durables, and their capital-values, can be readily excluded, as a direct determinant of W and r, from the definition of capital *per se*; treated, instead, as non-reproducible inputs. The argument, however, whatever its merits, seems inappropriate to the assumed context of stationary long-run equilibrium, where relative prices are constant over time. It does not appear in the *Lectures*.

12 This seems to imply that depreciation allowances can never be reallocated to the production of new (or different) capital goods. Wicksell does not explain why the conversion from fixed to circulating capital is 'usually impossible'. There is an implicit suggestion, however [Wicksell (1934, I:187)], that a change in the pattern of demands for final output, will imply that the rentals on some capital goods will no longer cover depreciation charges, i.e., that the value of some capital goods is less than their 'original cost', minus depreciation.

13 The appeal to adjustment-time—'where centuries are mere episodes'—in fact, contrary to Wicksell's recollection, in the above quote, does not appear in *VCR*.

14 Note Wicksell (1934, I:166): 'Some forms of capital (such as houses, railways, certain forms of improvements of land, etc.) normally last so long that the quantitative and qualitative adjustments, theoretically necessary for attaining equilibrium, become impossible in practice.' Later, in the review of Cassel, Wicksell, in fact, adopts the full adjustment point of view. See Appendix A below.

15 Note Wicksell (1934, I:193): 'If we imagine a society under more or less stationary conditions, in which a given capital in the possession of the propertied classes is employed year after year without appreciable increase or decrease, then each year about an equal part of that capital will be set free. That part (together with the consumable direct products of labour and land) constitutes the whole production of finished commodities and services of the year. When the capitalist class has taken the surplus, corresponding to interest on its capital, it must, in order to maintain its capital, re-invest the remainder—which it does by hiring labour and land for new production.' See Wicksell (1934, I:202–3, 207) as well.

16 To conform to current convention, some of Wicksell's notation is changed where convenient.

17 Wicksell (1954:22). See equations (4.1.17)–(4.1.20).

18 Note Wicksell (1954:120): 'Since we still wish to keep the hypothesis of the stationary condition and must consequently regard the sum of capitals as an invariable magnitude, we can disregard completely the gain which will ultimately result and which would obviously be a *new capital*.' The absence of an assumed stationary equilibrium, as Wicksell (1954:120n) sees it, is 'one of the fundamental mistakes in Böhm-Bawerk's presentation.'

19 Wicksell (1954:115). See equation (4.1.12).

20 In the informal discussion, Wicksell (1954:120, 121), at the firm-level, speaks of capital, not as a physical scalar $\bar{K}$, but as an amount 'borrowed', i.e., an 'amount of money on loan at a rate of interest which for the time being we shall think of as *given*.'

21 See equations (4.1.14), (4.1.22), (4.1.24) and (4.1.25). Wicksell (1954:124) assumes a uniform distribution of production processes, reading q as net output and does not explicitly write equation (4.2.4) until later [(1954:138)] in the argument. The approximation involved in the use of simple interest is also noted [(1954:123n)], i.e., equation (4.1.12) holds only 'for sufficiently small values of r and values of t which are not too great.'

22 Wicksell's (1954:122) graph of this function shows f(T) as strictly concave, i.e., $f'(T) > 0$, $f''(T) < 0$, with (apparently) $f'(T) = 0$ for some finite T. The latter leaves open the possibility that the rate of return to capital can be driven to zero. A formal production function is written [(1954:137)] later.

23 Wicksell (1954:126): 'If now, within the branch of the business in question, the total existing capital and the total number of workers employed were each a constant magnitude, we could find out not only the above relations between wage, level of interest and the length of the period of production (which, to be sure, can be assumed to be equal practically everywhere within this branch of the business), but even these magnitudes themselves.'

24 Note the correspondence to one-sector neoclassical theory: from $q = f(k)$; $W = f(k) - kf'(k)$ and $dW/dk = -kf''(k) > 0$ for $f''(k) < 0$.

25 Cf. equations (4.1.26)–(4.1.28).

26 Wicksell (1954:141–2): 'As is well known, Thünen had already laid down a law of the level of the rate of interest, analogous to his familiar proposition which stated that the average wage depended on the "yield of the last worker." According to this law, the level of the rate of interest depends on the productiveness of the "last invested particle of capital." The agreement of this theorem with Böhm-Bawerk's own is obvious and rightly emphasized by the latter. Only it must be remembered that here it is always a question of the capital investments of the individual entrepreneurs only, in which case the wage can and must be assumed to be given. This theorem can by no means be applied to the increase in the national capital itself and to the surplus return brought about thereby.' See Wicksell (1934, I:148) as well.

27 Note, from the Preface to *VCR*, Wicksell (1954:23–4): 'At the same time, I have succeeded in carrying through a *generalization* of the equations concerned. By this means the *services of land*, which Böhm-Bawerk did not take into consideration, are incorporated in the theory. Although this extension was pretty obvious, I may perhaps draw the attention of the interested reader to my formulae [equations (4.2.10)–(4.2.14)], which are relevant in this connection; for as I see it, the relationship between the main economic factors, labour, land and capital, is here for the first time exactly stated. This, under given assumptions, makes possible the theoretical determination of wages, rent and capital interest.'

28 See Cass (1973:91–6). With more than one 'circulating' capital good, Wicksell is clear that K must be measured in value terms and, as a result, cannot be taken as an exogenous variable—as if it stands 'outside' the equations characterizing a long-run equilibrium. See Sections 4.2c and 4.3 below.

29 Note that equations (4.1.1) and (4.1.3)–(4.1.5) are open, with (4.1.7) closing the system.

30 See Malinvaud (1972:240–41). For a further clarification of the distinction between constant- and current-value prices, see Appendix C.

31 The sole difference between one-sector neoclassical theory [see footnotes 2 and 24 above] and Wicksell's system, without land, is the extra variable T, implying the need for a savings function. In the former, k is treated as a given 'original' factor of production. If k does not depreciate, there is no room, without, for example, steady-state growth, to accommodate savings behaviour, i.e., $q = f(k)$, $q = W + rk$, $r = f'(k)$ and $k = \bar{k}$ are four equations in q, k, W and r. Letting all inputs grow at an exogenous rate g, in Wicksell's system, would make for a perfect analog to one-sector neoclassical growth. A 'Golden Rule' condition, or $r = \bar{g}$, generating maximum consumption per capita over time, closes the Wicksell system; also allowing for capital to be expressed as a 'subsistence fund' [equation (4.1.10)], continuous compounding notwithstanding.

Note, too, that the Wicksell system, with land, reduces to one-sector neoclassical theory as $T \rightarrow 0$; i.e., as production becomes instantaneous. Equations (4.2.10) and (4.2.11) become $q = f(l)$ and $q = W + Rl$, so that returns are generated to 'original' factors, land and labour, only. The marginal product or return to capital—as an intermediate good—is driven to zero; standard teaching in neoclassical theory: see Samuelson (1943) and, in particular, Lange (1936).

32 Assume that only capitalists save. Savings is also a function of relative prices over time. In a stationary economy, however, this effect is captured, in discrete terms, by r_t.

33 This result is confirmed by Negishi (1982b:172) in an overlapping-generations context. Note that with circulating capital alone, there is no confusion here (although it may appear so) between stocks and flows.

34 Note Wicksell (1934, I:205): 'If it were permissible to calculate with simple interest, the problem would be simplified in so far as the accumulation of capital through time need not be taken into consideration—though its distribution as labour-capital and land-capital, advanced wages and advanced rent, must; we should then only have to deal with the average period of investment.'

35 See fn. 8, Section 4.2a above. Wicksell (1954:109) also considers the 'second ground', i.e., 'the subjective and often *incorrect* underestimation of future wants resulting from defects of imagination and will', to be an issue regarding '*the formation of capital*'; one not appropriate to the assumed conditions of a stationary economy. See Wicksell (1934, I:154–5) as well.

36 Wicksell does not use the term 'rate of time preference', but simply expresses the idea in terms of the ratio of marginal utilities.

37 Wicksell (1954:108). Note Wicksell (1954:21): 'I think that Böhm-Bawerk's main formula for the explanation of interest—that interest is an agio or premium which arises from exchange between present and future goods—is quite correct and appropriate. But I cannot give it such a decisive significance compared with the older theories as was probably required by the author himself. Its applicability is shown, it seems to me, more in the theory of capital *formation* than in the real theory of interest.' See Wicksell (1954:112) and Wicksell (1934, I:169) as well.

38 The context of the following quote is discussed more fully in Section 4.3c below.

39 See Wicksell (1934, I:193). Wicksell's (1934, I:203) displeasure with the state of savings theory—'the laws of capital formation have been too little studied for a treatment of the subject in its entirety to be of much real use'—does not negate his recognition that a savings function is required for closure. See Section 4.3c below.

40 See Wicksell (1958 [1900]:110–1) as well.

41 See Wicksell (1934, I:169). Note Wicksell (1958 [1911]:183–4): 'Assume provisionally . . . that conditions are stationary and that there is a certain social capital which is maintained from year to year without increasing or decreasing, or, what amounts to the same thing, that a certain average *time* elapses between the application of productive forces and the appearance of the consumable products. It is then comparatively easy to show—at least now that Böhm-Bawerk has indicated the way—why and how this capital must earn interest at a certain unchanging rate under free competition. The *consumption period* and the *stock* of available present and future goods are then completely determined, since each year resembles the preceding one exactly, and every year *a whole "years' allocation"* of labour and land, composed in a definite given way of labour and land of different "ages" must be used. The difference in values between present and future goods (or past and present goods) which constitutes interest is therefore clearly determined by their different marginal utilities or marginal productivities—not by their total utilities, which do not enter into the comparison. The particular difficulty which Böhm-Bawerk and his readers must contend with, namely the fact that an increased *quantity* of goods does not necessarily imply increased *value*, does not exist here, because the exchange values of the different goods and services with respect to one another remain unchanged from year to year under stationary conditions.'

42 Note Wicksell (1954:115): 'Böhm-Bawerk's formula is thus undoubtedly the most general of all. It brings out better than the earlier explanations the true essence of the matter, namely the economic significance of *time*, and it adapts itself quite as well as any other mode of explanation to the different phenomena of interest.'

43 Wicksell certainly did not see, from equations (4.1.8)–(4.1.10), that r = g would recapture the 'subsistence fund' idea, even with compound interest.

44 This procedure creates an interesting problem for Wicksell, one that is prominent in the *Lectures*. Since ρ, in and of itself, is not assumed to be a positive constant, the equality ρ = r, although considered for analytical purposes to be a stationary equilibrium, does not, in fact, designate a steady-state. A value for r > 0, in other words, implies further capital accumulation, or net savings, and thus a fall in the rate of return to capital (and with it the rate of time preference) until

$$\rho = r = g = 0$$

obtaining a Classical stationary state. For a further discussion, see Appendix B below.

It is not unreasonable to conjecture that part of Wicksell's rationale for treating K *as if* it were exogenous—to separate-out savings behaviour—is explained by a desire to avoid this complication. The *raison d'être* for both *VCR* and the *Lectures* is to account for a positive rate of return to capital. Nevertheless, Wicksell's intuition for treating K as a physical scalar (with its one-good assumption, identical productivities across firms, and so on) *is* correct. As such, the focus is on K, an amount of corn, already accumulated.

45 Cf. Robinson (1969:396–7) on Wicksell.

46 Wicksell (1954:154): 'If the supply of, and demand for, the one commodity are equal, and consequently also equal with regard to the other commodity, equilibrium is attained; if not, a shifting of prices must take place. *But this change has obviously no influence on the proportion of production on both sides.*'

Note, also, as part of a more general setting, that two economy-specific savings functions are required to determine the size of K_1 and K_2. Although Wicksell does not make the point explicit here, he is clearly (as discussed above) aware of the need, in such a system, for this extension.

47 Equation (4.2.26) is written as in Wicksell (1954:160). Again, the importance of savings behaviour for closure is a point not explicitly made in this part of *VCR*. Nevertheless, it does appear—it is essential to note—in the *Lectures*, *and in the same analytical context*. See Section 4.3c below.

48 Wicksell (1954:160) correctly reasons that adding any number of additional consumer goods to the system simply adds as many new independent equations as unknowns.

49 Wicksell (1954:156): 'We can, however, determine without difficulty the position of equilibrium finally attained, with the help of our equations set forth above—but only if we assume that the present capital is a known magnitude.' The provisional nature of the solution to (4.2.29)–(4.2.33) is again noted.

50 Wicksell (1934, I:148). On the problems with simple interest, see Wicksell (1934, I:184), and on prices as parametric, see Wicksell (1934, I:103) and Section 4.3b below.

51 Wicksell (1934, I:145). For wages as a sum of goods, in exchange value, see Wicksell (1934, I:151, 187). Note Wicksell (1934, I:145): 'Since the yield of production is also measured in value terms, capital, like loaned money, has the peculiarity that its share in the product—interest—is the *same kind* of thing as capital itself; interest is the organic growth out of capital, a certain *percentage* of capital, whereas wages as against labour, and rent as against land, are quite heterogeneous things.' See Wicksell (1958 [1900]:107–10) as well.

52 See Chapter 3 above.

53 With simple interest, an approximate correspondence still exists between (4.3.1) and (4.3.2). See equation (4.1.11).

54 Wicksell (1934, I:149–50).

55 The return *r*K, is not given in corn, but, instead, *per*-Wicksell (1934, I:161): 'The surplus marginal productivity of all the groups of capital employed during the year, i.e., the total profit on capital of the year, constitutes *one year's interest* on the whole value of the total capital, each capital group being regarded as representing the value of labour and land employed, *together with the accrued interest*.'

56 See Wicksell (1934, I:168) and note Wicksell (1958 [1911]:178–9): 'Already in the introduction [to Böhm-Bawerk's *The Positive Theory of Capital*] comes the brilliant suggestion that the capitalist process of production—"the adoption of wisely chosen roundabout methods of production"—should be regarded as the *primary* concept, and capital itself as *secondary*: "essentially the intermediary products brought into existence at particular stages of the time-consuming roundabout process of production". This idea, which really renders superfluous all further discussion of the limits and extent of the concept of capital, is taken further in the masterly Book II, "Capital as an Instrument of Production." It is not completely rounded off, however, until the fourth and fifth sections of Book III, "The Origin and Rate of Interest," especially in the second main part of the latter, "The Rate of Interest in the Market." There, for the first time in the literature of this subject, a proper analysis of the relation between wages and interest is given, an entirely rational interpretation of the varying rate of interest under free competition—although it is true that the assumptions introduce very considerable simplifications, and that *land* as a factor of production is intentionally omitted. These parts of the book might as a matter of fact be read by themselves, as a complete and valuable whole. But Böhm-Bawerk has not been quite consistent here either: for in his account of the basis of the quantitative determination of the rate of interest on capital, he reverts, probably for didactic reasons, to the earlier Jevonian conception that national capital is a "Subsistence fund," an aggregate of (potential) wages, so that capital once more becomes the primary concept and the capitalist process of production to some extent a secondary, derived concept.'

57 Wicksell (1934, I:173): 'The whole vintage will be stored, either by the producers or by other entrepreneurs, for a number of years—in order that it may be sold to greater advantage. How *long* it will be stored depends, as we shall soon see, exclusively upon the amount of existing capital, which, on our assumption of a closed economy, can neither be increased by additions from outside nor diminished by export. The whole of the circulating capital of that society will consist of stored wine, though it can at any time be wholly or partially converted into money; we still make no definite assumption about the *value* of this capital in terms of money, but we assume that it just suffices for each year's vintage to be stored for a particular period (say four years).' The storage period, or the 'period of investment' τ, for Wicksell, is itself, also a variable. See Wicksell (1934, I:174) as well, where 'the amount of *capital* in the community, reckoned in terms of money', is listed among the 'unknowns of the problem'.

58 Since Wicksell (1934, I:174–5, 178) measures τ in years, the expression for r_τ, with continuous compounding over the course of a 'year', runs, in discrete form, in terms of inequalities. See equation (4.3.5) below.

The wine own-rate of interest is $\mathrm{r}_\tau = (\mathrm{p}_\tau - \mathrm{p}_{\tau+1})/\mathrm{p}_{\tau+1}$. Wicksell does not investigate the precise relationship between r_τ and r_τ. He is aware, however, that a $\mathrm{r}_\tau > 0$ implies rising current-value prices over time, as expressed in equation (4.3.3). See Section 4.2c above and Appendix C.

59 See equations (4.1.1)–(4.1.5).

60 Equation (4.3.17) is redundant to (4.3.15).

61 There is a related point. Since the above systems are properly closed, all the conditions, following Cass (1973:91–6), are in place—with a one-good assumption and constant-value prices—for q, rk, W and K to exist as real aggregates, even with r continuously compounded. The stock of capital, in particular, reads as a physical scalar, i.e., as so many gallons of wine. See Section 4.2b above. As such, the system (4.3.7), (4.3.9), (4.3.11) and (4.3.13) is the perfect continuous-time analog to the 'corn model' contained in *VCR*. *The main difference, of course, is that* $\mathrm{p}_\tau/\mathrm{p}_{\tau+1}$ *is not given in* (4.2.2)–(4.2.6), *leaving that system open*. That Wicksell chose in the *Lectures* not to represent K as a physical amount of wine—instead to work solely in value terms right from the outset—is undoubtedly due to the fact that, with continuous compounding, the measure of 'social capital' is not, as he sees it, independent of r_τ or prices. With simple interest, Wicksell, in *VCR*, treats both T, the average period of production, and K, as if they do not directly depend on the rate of return to capital; an option not available in the *Lectures*.

62 In the following, note the formal analog [see Section 4.2b above] to equations (4.2.10)–(4.2.14), with $\tau = \theta$, from *VCR*.

63 Cf., again, equations (4.2.10)–(4.2.14).

64 Wicksell (1934, I:172–3).

65 See equations (4.2.30)–(4.2.33).

66 Equations (4.3.33) are redundant to first-order conditions (4.3.30).

67 Indeed, this should not be surprising. The same structure of equations and unknowns characterizes both Sraffa (1960) and standard neoclassical models. For convenience, consider land to be a free good. Sraffa's (1960:34–5) price equations, reduced to 'dated quantities of labour'

$$p_jQ_j = WN_{0j} + WN_{1j}(1 + r) + WN_{2j}(1 + r)^2 + WN_{3j}(1 + r)^3 + \ldots$$

are identical to (4.3.33). Working forward to the final stage of production, replace N_{4j} and a produced capital good, with an amount of intermediate product K_{3j}, a period later; N_{3j} with K_{2j}; N_{2j} with K_{1j}, and so on. Summing each K_{tj}, over all $t > 1$, gives an aggregate amount i of each capital good K_{ij} or Q_{ij}, to produce Q_j, and price equations

$$p_jQ_j = (1 + r)\,[p_1Q_{1j} + p_2Q_{2j}] + WN_j.$$

Sraffa fixes Q_j, Q_{ij} and N_j exogenously, so that setting $p_1 \equiv 1$, for two goods indexed by j, gives two equations in three unknowns: p_2, W and r.

In neoclassical terms, index $k_{ij} = K_{ij}/Q_j$ as the amount of each capital good i used to produce per-unit j. Per-unit price equations, for $n_j = N_j/Q_j$, are

$$p_j = (1 + r)[p_1k_{1j} + p_2k_{2j}] + Wn_j$$

with 'optimal' input coefficients, k^*_{ij}, from first-order conditions, giving two equations in three unknowns: $\pi = p_1/p_2$, W and r. The value of capital reads simply as $K = \pi K_1 + K_2$.

68 In the secondary literature, e.g., Petri (1978:252), the following quote is used as evidence of an admission, on Wicksell's part, of the error involved in taking the value of capital as exogenous. In fact, however, Wicksell is clear, throughout *VCR* and the *Lectures*, that the measure of K *must* be endogenous to equilibrium prices, whenever more than one 'circulating' capital good, say, corn and linen—or a single 'circulating' capital good defined, for example, as the value of the stock of wine across successive vintages—is being considered.

69 Note, for Wicksell, that with simple interest, 'circulating' capital, and an average period of production T, *the amount of savings is the stock of capital*. See Section 4.2c above.

70 It is worth noting that part of the explanation for why savings behaviour is not more prominent in Wicksell's *formal analysis*, must be due to Wicksell's (1934, I:217) perception of the state of savings theory as 'incomplete', as 'too little investigated'; e.g., Wicksell (1934, I:203): 'In reality, the amount of capital is not determined by physical conditions, but by the equilibrium between psychical forces which, on the one hand, drive us to accumulate capital and, on the other, to consume already existing capital. In other words, *the accumulation of capital* is itself, even under stationary conditions, a necessary element in the theory of production and exchange. We have now reached a point in our exposition at which this new factor forces itself upon our attention. We shall, therefore, consider this subject in our next chapter [Part III of the *Lectures*]—though the laws of capital formation have been too little studied for a treatment of the subject in its entirety to be of much real use.' And later, Wicksell (1934, I:207–8): 'Unfortunately, such a theory [a savings theory] has not been worked out, and the phenomena which it should explain depend on a number of motives—partly selfish, partly altruistic, but in any case very complex.' Although Wicksell is obviously aware that a savings function is required to close his long-run systems, his [Wicksell (1934, I:217)] own discussion of savings behaviour is designed merely 'to indicate, rather than solve, the many problems associated with the question of capital accumulation'. The conditions, put simply, for stationary equilibrium, i.e., net savings equal to zero, a resulting amount of K, constant income and relative prices over time, etc., are easily set out. The determinants, however, for additions to the stock of capital—for how the economy arrives at a given stock, 'already accumulated'—are not. Since a full-fledged theory of savings 'has not been worked out,' setting S(r) = 0 becomes, in the interim, the simplest way to proceed.

Chapter 5

Closing remarks

What lesson can be gained from the preceding studies? At first glance, especially if one has in mind the standard accounts contained in the secondary literature, the long-run systems of Walras and Wicksell appear quite different, if not incongruous. The idea, for example, of a 'period of production', reinforced by all of the Austrian paraphernalia contained in Wicksell's capital theory, seems to have little in common with the 'static' equations which characterize Walrasian general equilibrium. Indeed, at least until Part VI of the *Elements*, where the role of circulating capital is explicitly considered, Walras (1954:242, 282) completely ignores the length of time between the application of inputs and the realization of output, assuming instead that production and exchange take place instantaneously.

In fact, however, what the preceding studies demonstrate is a clear equivalence, an 'essential continuity' in method and content between the long-run theories of Walras and Wicksell.[1] Both were preeminently interested in the structure of equations and unknowns that define a stationary long-run equilibrium; a system where net income, consumption, relative prices and the measure of the capital stock all remain constant over time. Both constructed a fully coherent and consistent long-run theory, a framework for determining a positive and uniform rate of return on capital in a supply and demand context, without the need to assume either steady-state growth or static price-level expectations, and without the need to treat the aggregate value of capital as an exogenous variable.[2] Both closed their long-run systems (although, as discussed in Chapter 3 above, Walras was not properly explicit on this point), with an assumption of zero net savings, leaving the total value of capital, most importantly, endogenous to equilibrium prices and quantities, as well as to savings behaviour.[3]

There is, in addition, clear evidence that Wicksell (near the end of his career) came to realize this 'essential continuity', i.e., that he came to realize the direct correspondence between his own system and that of Walras. The recognition occurs in two related areas: (a) in Wicksell's re-evaluation of Walras's capital theory, from highly critical remarks to a more sympathetic understanding; and (b) in Wicksell's eventual acknowledgement that his own 'short-period', long-run theory could be extended to a 'full' long-run equilibrium [see Section 4.2a above], where fixed capital goods are included—a system, for all essential purposes, identical to that contained in Part V of the *Elements*.[4]

My view of Walras and Wicksell has important implications for the history of late nineteenth century economic thought. The main problem with standard interpretations, as I see it, is the common belief—with great injustice to the historical Walras and Wicksell—that one can view all of stationary long-run theory in this period solely in terms of a Clark-like construction.

Garegnani (1976) provides a good example of what I have in mind. Contrary to the typical neoclassical approach of Hicks (1946) and Hirshleifer (1967) that I outlined in Chapter 1 above, Garegnani (1976:34), as we have seen, charges that Clark, Böhm-Bawerk, Wicksell and Marshall all take the erroneous position of assuming 'the amount of capital in the economy to be a given (value) magnitude and not an unknown determined by the condition of zero net saving.'

The 'only exception' to this is thought to be Walras, who treats the initial amounts of each capital good as arbitrarily fixed and given. But since, according to Garegnani, Walras attempts to construct a long-run equilibrium with uniform returns, he simply errs in this regard. With a capital endowment set in these terms, Walrasian theory can only give a short-run position, where the rental on each capital good, as a fraction of its supply price, will generally vary throughout.[5]

Moreover, with Walras's apparent inability to construct a theory of long-run equilibrium in mind, Garegnani (1976:34) goes on to claim that late nineteenth century long-run theorists in fact 'had no alternative' to treating the aggregate value of capital as a datum:

Marginal theorists [Wicksell, Clark, Böhm-Bawerk, etc.] not only generally conceived the capital endowment of the economy as a value magnitude, but also had no alternative to doing so if they were to be consistent with the conditions of long-period equilibrium; that 'full equilibrium' in which, to paraphrase Hicks (1932), the marginal product of labour is the one resulting when the 'quantity', but not the 'form', of the 'co-operating capital' remains unchanged.

The theory, in other words, purportedly requires a given measure of the aggregate value of capital—allowing the relative amounts of each capital good or the 'form' of capital to change until returns are uniform—to enter as a parameter in the demand function for labour, thus determining the wage rate, relative to the (homogenous) stock of labour.[6]

Now obviously Garegnani's account of this period, if valid, is quite damaging. If a knowledge of the aggregate value of capital is needed prior to the determination of equilibrium returns, the theory is inconsistent from the very outset. Although Walras, as the 'only exception', does not employ the assumption of taking the aggregate value of capital as a datum—starting instead by treating each capital good as arbitrarily fixed and given in physical terms—the best that he can obtain, as a result, is a short-run equilibrium. In the history of late nineteenth century economic thought, in other words, there is simply no adequate or consistent theory of a stationary long-run equilibrium.[7]

But this is not true. What I have shown in the preceding studies is that nothing in Garegnani's account of late nineteenth century theory applies to the historical Walras and Wicksell. Indeed, not only is there an adequate theory for this period, but, as I indicated earlier in this chapter, both Walras and Wicksell set out what amounts to the same theory of long-run equilibrium for a stationary economy. There is no history of error here. Wicksell does not take the aggregate value of capital as a datum and Walras is clearly not limited to a short run theory.[8] Nothing in their argument, including the construction of a demand function for labour, requires a prior measure of the value of stock of capital.

What Garegnani sees as characteristic of all of late nineteenth century stationary long-run theory—a criticism, as we have seen, that applies just as well to the view of Wicksell found in Hirshleifer (1967), Lutz (1968), Samuelson (1982), etc.—should in fact only be attributed to Clark (1965 [1899]). It is Clark, in other words, who treats the '*fixed amount of social capital*' as a parameter in the demand function for homogeneous labour. It is Clark who takes the aggregate value of capital, an 'amount of "money" permanently invested in a succession of material things', as a datum.[9]

The claim that late nineteenth-century theory 'had no alternative' to this is simply incorrect. Wicksell and Walras are not 'marginal theorists' in the sense of Clark. Both close their systems of long-run equations with a savings function and an equality between (zero) net savings and investment, leaving the measure of value-capital endogenous to relative prices and the savings behaviour that defines a stationary equilibrium; essentially the same (consistent) procedure used by Hicks (1946) and Hirshleifer (1967).

Notes to Chapter 5

1 As I show in Appendix C below, this 'essential continuity' carries forward through Hicks and the Neo-Walrasian idea of an intertemporal equilibrium.

2 Some care is required here. The Part IV system in the *Elements* is also a stationary long-run equilibrium, but one without a uniform rate of return. Each capital good is treated on the same footing as land or labour, i.e., as a non-produced or 'original' factor of production, given in arbitrary amounts. See Section 3.2 above. My observation in the text refers to the Part V stationary equilibrium, where Walras now treats capital goods as re-producible inputs.

3 The assumption of stationary conditions, in effect, deprives the time-element in Wicksell's 'period of production' of much of its substantive importance. By assuming a uniform distribution of production processes, all with the same period of production τ, Wicksell obtains a stationary flow of net output and constant relative prices over time. As long as capitalists continuously replace the stock of capital (saving and re-investing all of net returns), the value of capital will also remain unchanged, earning a uniform rate of return. This is the reason why Wicksell's systems—those with only circulating capital—can be so easily compared to standard one-sector neoclassical theory. See footnotes 2, 24 and 31 in Chapter 4 above.

4 See Appendix A below for the relevant exegesis and a further discussion.

5 As I noted in Chapter 1 above, Petri (1978) and Robinson (1961) interpret Walras in a similar manner, with Robinson's position as the most extreme, i.e., here, the best that a Walrasian theory of supply and demand can consistently provide is a Marshallian short-run equilibrium, where both the aggregate and firm-specific amounts of each capital good are arbitrarily fixed and given.

6 A similar procedure, so this story goes, is simultaneously applied to the demand for value-capital and land, with the aggregate supply of value-capital as an exogenous variable, needed to determine the uniform rate of return and land-rent.

In Appendix C, I show that Garegnani also misinterprets Hicks (1932) as treating the value of aggregate capital as a datum, rather than as an unknown determined by a condition of zero net savings.

7 Garegnani's interpretation of Wicksell and Walras cannot be dismissed as a particular or doctrinaire view. As I detail throughout Chapters 3 and 4 above, what might (normally) be viewed as more orthodox or neoclassical versions of this period, argue much the same: Although it is not seen as a requirement of the theory, Wicksell is invariably portrayed as taking the value of aggregate capital as a datum; and although Walras, in these accounts, is not represented as a strictly short-run theorist, he nevertheless errs in his construction of a long-run equilibrium. Here, the *Elements* must be *re*-interpreted as a model of either steady-state growth or one in which static price-level expectations give a sequence of non-steady-state growth periods over time. The final outcome remains unchanged, i.e., even from a more orthodox point of view, there is still no adequate theory of a *stationary* long-run equilibrium for this period.

8 In fact, as discussed in Chapters 2 and 3 above, the idea of a short-run equilibrium cannot be found in the *Elements*. The Part IV system is essentially equivalent to a long-run, one-sector neoclassical theory of distribution; the use of 'tickets' in the *tâtonnement* for the 'Theory of Production' is specifically designed to exclude a Marshallian short-run, with equilibrium economic profits and losses, by characterizing, instead, adjustments around a given and invariant long-run position; and although, in Part V, the aggregate endowment of each capital good is initially given, in arbitrary amounts, net accumulation in the *tâtonnement* for 'Capital Formation' changes their prices, rentals, and relative quantities until a stationary long-run equilibrium with uniform returns is obtained. Again, contrary to Schumpeter (1954), Hicks (1934), Morishima (1980), etc., there is no need to impose the assumptions of steady-state growth or static price-level expectations on the *Elements*.

9 See Clark (1965:116–25, 157, 160, 197), and note the uncomfortable circularity in his discussion of uniform returns: 'It will be noticed that, in computing the rate of interest, we first ascertained the absolute amounts, or lump sums, earned by all the several instruments. In a sense, interest depends on rent: it is total rent, reduced to a percentage of total capital. In another and deeper sense, rent is governed by interest: the amount that any one instrument earns depends on the number of such instruments in use.' And later: '*Proximately, rent fixes interest*. Given a certain number of capital goods of each kind, and what they earn is the amount that, by an arithmetical reduction, is converted into interest. *Fundamentally, interest governs rents*. Given a certain permanent fund of capital, and it is put into such forms that the rent secured by one concrete form, or capital good, is as large a fraction of its value as is that secured by another.'

Appendix A

In this appendix, I provide the evidence for Wicksell's recognition of an 'essential continuity' in the history of long-run theory. As I claimed in Chapter 5 above, Wicksell eventually came to realize that his own system was clearly compatible with that of Walras.

To see this, begin with *VCR*, where Wicksell's long-run theory generates uniform returns on 'circulating' capital only.[1] Here, for the most part, the criticism of Walras centers on the characteristics of Part IV of the *Elements* as a stationary long-run equilibrium—where each input, including each fixed capital good, is treated as an 'original' factor of production:

> Walras calls 'capital' and treats as 'capital' *only* durable goods, but not raw materials and half-finished products and not the means of subsistence of workers. . . . In this interpretation the true role of capital in production is completely overlooked. A necessary consequence of this is the peculiar fact that these equations of production and exchange *can give no information at all about the level of the rate of interest*. If only durable goods are regarded as capital, then a certain rent is fixed for each group of these by the above mentioned equations, but *not* the *capital value* of the goods itself, nor, consequently, the *rate* of interest either, '*le taux du revenue net*'. This is explicitly admitted by Walras; but he asserts that, in order to determine the level of the rate of interest, it is necessary to turn from the investigation of a stationary economy to the investigation of a progressive one, where *new* interest-bearing capital goods are produced, whose capital value can be determined from the production costs. This is certainly incorrect. In the stationary economy, too—even if we assume that all means of production are indestructible—a rate of interest of the circulating capital will undoubtedly establish itself, precisely because the lengthier methods of production prove more profitable. (1954:167)

Without 'circulating' capital in the Part IV system, in other words, it is impossible—Wicksell is correct in all of this—to determine input prices or the rate of return. Each fixed capital good, given in constant and arbitrary amounts, simply earns an input-specific rental.[2] (As Wicksell also acknowledges, this is standard teaching in the *Elements*; points with which Walras would not disagree. See Sections 3.2 and 3.3a above.) The only way, as a result, to generate both 'capital values' and uniform returns (again, a point with which Walras would not disagree), is through net accumulation or by postulating a savings function with net additions to the stock of capital.

Note, however, that Wicksell, unlike my interpretation in Chapter 3 above, does not attribute the assumption of zero net savings to Walras. Although net accumulation is clearly required to 'move' from Parts IV to V of the *Elements*, a necessary characteristic of the *tâtonnement* for capital formation [see Section 3.4b above], Wicksell fails to indicate that once equilibrium is obtained in Part V, Walras's formal equations (as well as his more informal discussion) designate a stationary outcome.

In fact, in the *Lectures*, with sole reference to Part V of the *Elements*, Wicksell, while retaining the criticism regarding the absence of the 'marginal productivity of time', simply imposes the condition of zero net savings on the Walrasian system:[3]

> Walras's formula for interest . . . reduces itself, on the assumption of *stationary* conditions, simply to the equation S(r) = 0, in which S(r) is the amount of annual savings conceived as a function of the rate of interest r. In other words, it expresses the truism that, in the stationary state, the inducement to new savings must have ceased; but it affords no answer to the question why a given amount of existing social capital gives rise to a certain rate of interest, neither higher nor lower. The importance of the time-element in production was never properly appreciated by Walras and his school. The idea of a *period* of production or of capital-investment does not, as we have said, exist in Walras-Pareto theory; in it capital and interest rank equally with land and rent; in other words, it remains a theory of production under essentially non-capitalistic conditions, even though the existence of durable, but apparently indestructible instruments, is taken into account. (1934, I:171)

This too is a legitimate account of Walras. Until Part VII of the *Elements*, i.e., until Walras drops the assumption of fixed coefficients and establishes a theory for the marginal product of capital, he merely assumes, without explanation, that net rentals are positive. Walras's Part V system, with S(r) = 0, as such, lacks a productivity criterion; an equation (or set of equations and unknowns) to determine the rate of return, *per*-Wicksell, in terms of the period of production τ on circulating capital.[4]

As yet, however, there is still no suggestion by Wicksell that the equations for long-run equilibrium can be broadened to include fixed capital goods and, if so done, that the resulting system is quite compatible with Walras's Part V equations. This comes, finally, in the review of Cassel:

> Walras proceeds in an entirely different manner. For him the capital-*goods* themselves are factors of production just as much as labour and the forces of nature; and the rate of interest '*le taux du revenue net*' is considered as the ratio between the expected yield of the capital goods now being made (= the price for their factors of production minus the necessary amortization costs) and their own cost of production according to *present* prices. Thus it represents a 'parameter' in the 'functions' which determine saving. Savers and entrepreneurs strive to maximize this ratio and equilibrium is reached when it is the same for all alike. In this way Walras constructs an extraordinarily coherent and rigorous system, which, when it is combined with the systems of Jevons and Böhm-Bawerk, both completes and is completed by them. (1934, I [1919]:236)

And, in the footnote accompanying this passage:

> At this point I must withdraw an objection which I previously made against Walras—i.e. that his theory of interest necessarily presupposes a *progressive* type of society. Walras indeed said so himself, but the truth of the matter is that it is just as applicable to the stationary state, and in fact gains thereby in rigor. The underlying assumption is that the factors of production will have the same relative values or prices in the future as they have at the present moment. (1934, I:226n)

For i different kinds of capital goods, in other words, a 'full' long-run equilibrium gives (when savings and investment decisions are equalized) a uniform rate of return r_t defined by

$$r_t = (R^i_{t+1} - \delta^i)/P^i_t \qquad \text{(A.1)}$$

for P^i_t the undiscounted price, R^i_{t+1} the gross rental, and δ^i the corresponding rate of depreciation for each capital good, *just as in Walras's* (1954:268–9) *Part V equations*.[5]

Indeed, for Wicksell, the combination of the savings function and definition of uniform returns given by equation (A.1), found in Walras's 'extraordinarily coherent and rigorous system', together with Böhm-Bawerk's productivity criteria [or $f'(\tau)/f(\tau)$] for variations in the period of production, '*completes*' both systems.[6] The result is a unified stationary long-run theory—the 'essential continuity'—which can accommodate all types of capital goods, fixed and circulating. (Unfortunately, Wicksell does not recognize that the assumption that 'factors of production will have the same relative values or prices in the future as they have at the present moment' *is* a characteristic of Walras's Part V equations.) The only additional requirement is that net savings must equal zero; a condition which is repeatedly stressed throughout *VCR* and the *Lectures*, and which, as I have argued in Chapter 3 above, is part and parcel of Walras's theory of long-run equilibrium.

Notes to Appendix A

1 Recall that throughout *VCR* the stock of each durable capital good (given in arbitrary amounts) remains constant or intact with production. See Sections 4.2a and 4.2b above.

2 It is interesting to note that if Walras would have included 'circulating' capital, freely transferable between uses and earning uniform returns, in Part IV—rather than deferring its introduction to Part VI of the *Elements*—the resulting system would have represented a perfect analog to the theory contained in *VCR*.

3 Note Wicksell (1958 [1909]:172): 'I could not stand as a champion for every detail of Böhm-Bawerk's edifice of theory; but in order to convince oneself that it represents a step forward from the older ideas to which Pareto is still faithful, it is in my opinion only necessary to consider the formula which Walras . . . put forward as *his* solution of the problem of interest, as against Böhm-Bawerk. If conditions are assumed to be *stationary*, Walras's seemingly so meaningful formula reduces itself to the simple relation: $S(r) = 0$, where $S(r)$ is the total of new savings as a function of the rate of interest, r. In other words, the equation now says that under such circumstances the incentive to construction of new capital must have disappeared. This is correct, but insufficient, for we also want to know why and how a given total amount of capital warrants a certain definite rate of interest. We can find the answer to that question only by following a particular line of thought: by considering *the effect of time* on production and distribution, by considering the effect of a virtual lengthening of the production process.'

4 Although Walras (1954:45–6) is evidently unwilling to accept the idea of a period of production, he is clearly aware of a comparable defect (i.e., the absence of a productivity theory) in the Part V equations. See Part VII of the *Elements* and fn. 11, Chapter 3 above.

5 The term δ^i can accommodate all types of capital goods with varying durabilities, e.g., $\delta = 1$ for 'circulating' and $\delta = 0$ for 'perpetual' or non-depreciating fixed capital. For simplicity, Walras assumes a uniform rate of depreciation in Part V of the *Elements*, but then extends the argument in Part IV to account for various durabilities, including 'circulating' capital goods.

A hint of this long-run condition, in fact, comes in the closing pages of *VCR*—in the context of a 'dynamic equilibrium'—where Wicksell (1954:165n1) argues: 'The production of new rent-goods, for instance, must then be treated in the same way as the production of consumable goods, in which case, however, the sum of the circulating capital no longer remains unchanged. Instead, the condition is added that the newly produced rent-goods must yield as rent the usual capital-interest on the costs of production.'

6 Note Wicksell (1958 [1897]:146): 'E. Barone has recently shown in a really brilliant manner how Walras's theory of production may be completed, so that all theoretical blemishes are removed. This is accomplished simply by adding another relation to the 'logical relations' formulated by Walras for the quantities concerned; that is to say, by relating them to the *length of the production period* or the investment period, thus bringing into the foreground this exceedingly fruitful idea, which was indicated by Jevons, but first exploited to the full by Böhm-Bawerk.'

Appendix B

Although not essential to the argument of Chapter 4, there is an important issue that remains to be discussed, one with particular relevance to Wicksell.[1] I have in mind, in this appendix, the characteristics of the growth-path that result from 'stringing' together, with variations in the amount of capital, different long-run equilibria.

As detailed in Section 4.2c above, Wicksell is generally content with an equality between the rate of return (r) and the rate of time preference (ρ) as a condition, in 'preliminary' terms, for a stationary equilibrium. A positive rate of return in long-run equilibrium is explained by the relative scarcity of capital, say K*, measured as time-augmented dated labour- and land-inputs. Moreover, without the condition that $f'(\tau)/f(\tau) = r > 0$, constant income over time, for Wicksell, implies that $\rho = 0$, with no tendency for net savers to predominate.[2] A positive rate of time preference, in other words, requires a positive equilibrium rate of return to capital.

The problem, however, is that the condition $\rho = r$, in a broader theoretical context and, in particular, as Wicksell defines the rate of time preference (with constant consumption profiles), does not, in fact, constitute a stationary state. Since ρ is not restricted, *ex ante*, to be positive, a positive rate of return must imply the presence of net savings until $\rho = r = 0$, i.e., since capital goods are pure intermediate products, net accumulation will continue—without an assumed positive rate of time preference to stop the process—until net marginal products are driven to zero.[3]

Wicksell (1934, I:209–18) indeed sees this and, as a consequence, draws a distinction between a 'preliminary' stationary equilibrium with $r > 0$, and a 'completely' stationary state where the stock of capital, say K^{**}, is sufficiently large to generate a zero rate of return, so that $f'(\tau)/f(\tau)$, or the 'marginal productivity to time', equals zero:

> The progressive accumulation of capital must be regarded as economical so long as any rate of interest, however low, exists. For the average individual, or rather for society as a whole (regarded as an individual who never dies), the accumulation of capital presupposes the exchange of a lower marginal utility for a higher . . . Under such conditions, we should therefore expect a continual accumulation of capital—though at a diminishing rate—and, at the same time, a continual fall in the rate of interest. (1934, I:209)

From 'Marginal Productivity Theory as a Basis of Distribution in Economics':

> If we look at a society as a whole with an almost stationary economy, we can see that complete economic equilibrium can never occur, and the accumulation of capital can therefore never cease until the rate of interest falls to zero, which implies that the annual product has reached the maximum physically possible. For as long as stored-up labour and land provide even the smallest excess of marginal productivity over that of current labour and land, a sacrifice on the part of the average individual now—if it is not altogether too large—must give a larger compensation in the future. (1958 [1900]:118)

And, in the review of Cassel:

> One cannot with certainty infer any other conclusion than that saving and capital accumulation will progress at a slower *tempo* the more the rate of interest falls. And this seems to be clear *a priori*. Assuming a sufficiently clear insight into the urgency of future wants as compared with present wants, and also a sufficiently vivid interest in the welfare of future generations, it will appear that capital accumulation cannot cease, as long as it is generally possible to gain *more* in the future by sacrificing *less* (computed in subjective values) in the present, i.e. as long as there is a positive rate of interest, however small . . . the rate of interest would therefore tend to fall to a minimum, until it finally became zero. (1934, I [1919]:241)

At K^{**}, capital reaches its 'maximum importance' (as $f'(\tau)$ approaches zero), with returns accruing only to 'original' factors of production, labour and land.[4] The value of produced output is simply decomposed into total wage and rental payments.

Where does this leave Wicksell? In the systems of equations

where the absolute amount of labour is fixed exogenously, with a zero rate of growth, the resulting 'marginal product of time' is positive only if capital is scarce, i.e., if $K^* < K^{**}$. An equality between ρ and r defines a 'preliminary' stationary long-run equilibrium (with uniform rates of return and constant relative prices over time) as long as Wicksell excludes the possibility of any further additions to the total number of inputs; abstracting, in particular, from capital accumulation.[5] With net savings equal to zero, the stock of current and time-augmented labour and land, including its technical composition over τ, remains intact.[6]

If the argument is now extended to cover net additions to inputs, with positive net savings 'moving' the economy from K^* to K^{**}, the only steady-state—Wicksell's 'complete' stationary equilibrium—occurs when

$$\rho = r = g = 0 \qquad \text{(B.1)}$$

since the rate of growth of land, as a fixed input, is zero.[7] In fact, equation (B.1) is consistent with both: (a) capital attaining its 'maximum importance'; and (b) a zero own-rate of interest defined for each good taken separately, or

$$\rho_t = r_t = (p_t - p_{t+1})/p_{t+1} = 0 \qquad \text{(B.2)}$$

in constant-value prices, so that $p_t = p_{t+1}$.[8] With r = 0, a constant value for net income, hence, consumption over time (at its maximum for each t), implies a zero rate of time preference. As the rate of return falls with accumulation so does ρ, with the following implication. Although both the 'preliminary' and 'complete' versions of long-run equilibrium, as constructed, imply constant relative prices over time, it is clear that the 'move' from K^* to K^{**} must result in a change in prices intertemporally. Tracing out, in other words, a series of 'preliminary' long-run equilibrium states, as a traverse (questions of stability aside) to the steady-state defined by equation (B.1), gives a sequence for Q, W, R, p_j, r and so on, over t. A 'process', with capital gains or losses on durables, not unlike that defined by a Neo-Walrasian intertemporal equilibrium.[9] On this issue, however, Wicksell, at least in his theory of capital, had little to say.[10]

Notes to Appendix B

1 Although Walras, like Wicksell, does discuss the comparative-statics of a change in the stock of capital [see fn. 47, Chapter 3 above], he does not detail (and certainly not to the extent that Wicksell does) the resulting growth-path.

2 Note Wicksell (1934, I:208n1): 'Cassel is not quite correct when he says: "A man who attaches the same importance to future needs as to present ones, if he expects to be able to provide for his needs in the future just as easily as he does now, has no reason for setting aside anything of his present income." This argument actually presupposes the absence of any rate of interest.' See Sections 4.2a and 4.2c above.

3 This is standard neoclassical teaching, see Schumpeter (1961 [1934]:25–43) and Lange (1936). Although a rate of time preference forms no part of the argument contained in the *Elements*, Walras does suggest [see fn. 48, Chapter 3 above], like Wicksell, that a positive rate of return determines a positive ρ. Unlike Wicksell, however, Walras is apparently unaware of the fact that such a view must designate (again, in a broader theoretical context) that the Part V equations cannot be stationary.

4 Wicksell (1934, I:214). Wicksell (1934, I:211–4) is careful to list a variety of 'checks' to the tendency for $r \to 0$: (a) the 'second ground' for the existence of interest, i.e., 'the subjective and often *incorrect* under-estimation of future wants'; (b) technological change; (c) the occasional 'destruction of capital', from war, accident, the business cycle, etc.; and (d) foreign investment. Cf. Mill (1965, III [1871]:741–46).

An increase in capital, for Wicksell (1934, I:157), also increases the 'marginal productivity of (direct) labour and land'. With rents rising and the rate of return falling, the price of land, i.e., $P_L = R/r$, as an argument in $S(\cdot)$, increases with accumulation, until it possesses an 'almost infinite exchange value'. Holders of claims to land (as a source of 'perpetual income') experience capital-gains over time; with, as Wicksell (1934, I:214–15) notes, a 'large part' of 'apparent annual savings' accounted for by an increase 'in the capital value of land'.

5 Wicksell (1934, I:164, 211, 213) also extends the argument to cover, setting land aside, equal increases in the amount of labour and capital, but fails to write $\rho = r - g$, for $g = \dot{N}/N = \dot{K}/K$, as an equilibrium statement, e.g., Wicksell (1934, I:213): 'If the growth of population is accompanied by an increased demand for all kinds of products, on the one hand, and by an increased supply of labour available in the future, on the other, then a capital accumulation which might have brought down the rate of interest to practically nothing under stationary conditions will not now be sufficient to do so; or will only just suffice to maintain capital at about the same *relative* level, for which reason it will continue to possess a high marginal productivity and to yield a high rate of interest.'

6 Note, however, that given Wicksell's definition of ρ, an equilibrium with $r = \rho > 0$ must also designate that consumption profiles—

one of the principal characteristics of a stationary system for Wicksell [see Section 4.2a above]—cannot be constant over time. Wicksell, in taking the value of capital as provisionally given in these kinds of models, seems to have missed this point altogether.

7 As Wicksell (1934, I [1919]:226n1) remarks: 'The underlying assumption [of a 'preliminary' long-run equilibrium] is that the factors of production will have the same relative values or prices in the future as they have at the present moment. Actually this is true for the stationary state, but it does not hold for the progressive economy unless we postulate a *uniform* increase in production, which is strictly speaking inconceivable, as the sum of natural forces cannot be increased.'

All of this is clearly reminiscent of Classical theory. See Hollander (1979, 1987).

8 See Section 4.2c above. In this case, the measure of r is the same regardless of whether it is expressed in constant- or current-value prices.

9 See Appendix C below.

10 Cf. Wicksell's (1935, II [1911]; (1965) [1898]) well-known distinction between a 'market' and 'real' rate of interest.

Appendix C

To further establish the thesis of an 'essential continuity' in the theory of long-run equilibrium, this appendix gives a brief sketch of the progress in long-run theory after Walras and Wicksell. In particular, I have in mind the work of Hicks and the development, based largely on Hicks's approach, of the Neo-Walrasian idea of an intertemporal equilibrium, especially since these systems are typically viewed [e.g., Milgate (1979)] as so radically different from late-nineteenth century neoclassical theory.

In the *Theory of Wages*, Hicks (1932) adopts a standard neoclassical method. The principal concern is with the theory of marginal productivity, especially as it applies to labour. Complete substitutability is assumed and the resulting equilibrium conditions generate the familiar equality (with a capital structure that is fully adjusted to its long-run outcome) between the wage rate and the value of the marginal product of labour.[1] The value of the aggregate stock of capital is endogenous to relative prices and savings behaviour. The long-run equilibrium contained in the *Theory of Wages*, in other words, is essentially identical to the long-run theory of Walras and Wicksell.[2]

This is an important point. Garegnani (1976:33) incorrectly interprets Hicks (1932) as employing a 'Wicksell approach' of fixing the value of aggregate capital—not an unknown determined by the condition of zero net savings—to close the long-run system. But this, as we have seen, is generally *not* the case for Wicksell (or, for Walras) and, indeed, this is not the way in which Hicks sees the stationary state idea.[3] As he explains in *Value and Capital*:

How will the quantity of intermediate products—the quantity of capital—be determined?

It turns out to be determined through the rate of interest. A fall in the rate of interest would encourage the adoption of longer processes, requiring the use (at any moment) of larger quantities of intermediate products. But since we are in the stationary state, there can be no tendency for the stock of capital to increase or diminish; constancy of the stock thus gives us one relation between its size and the rate of interest. Also, if entrepreneurs do not desire to increase or diminish their stock, their net borrowing must be nil. If the demand and supply for loans are to be in equilibrium, net saving must therefore also be nil. The rate of interest must therefore be fixed at a level which offers no incentive for net saving or dis-saving. What this level is depends partly upon the propensities to save of the individuals composing the community, partly upon their real incomes—and these depend again on the size of the stock of intermediate products. We therefore have two equations to determine the size of the stock of capital goods and the rate of interest; consequently both are determined. (1946:188–9)

This is the procedure used in *The Theory of Wages*. The stationary state is a point where the equilibrium amounts of net savings and investment are equal to zero.[4] The value of aggregate capital is not taken as a datum.

In fact, it is not a problem with capital—in the stationary state its equilibrium amount is perfectly determinate—that resulted in Hicks's movement away from the approach in *The Theory of Wages*, but the static and stationary conception of long-run equilibrium that it contains.[5] He (1963:307) dates his 'change in emphasis', his 'own personal "revolution" to May or June 1933.' The date coincides with the influence of a paper by Hayek (1928) on 'intertemporal equilibrium' and Hicks's desire to construct a 'dynamics' that improves upon the comparative-statics method of, first, assuming that a long-run position exists and then, second, 'stringing' such equilibria together as a suggestion of some sort of temporal process.[6] To do so, for Hicks (1965:24), requires introducing 'expectations of the future, of the future after time t, if the equilibrium values at time t are to be properly determined'.

Value and Capital makes the nature of Hicks's personal 'revolution' especially clear. He (1946:115) begins by defining 'statics' as 'those parts of economic theory where we do not trouble about dating,' and 'dynamics' as 'those parts where every quantity must be dated.' Static theory, in turn, could only be used to study a temporal process, Hicks (1946:116) continues, 'if we could say that the system of prices existing at any moment depends upon the preferences and resources existing at that moment and upon nothing else'. But, generally, this is not the case:

The main reason why it is not true is that the adjustments needed to bring about equilibrium take time. A rise in the price of a commodity exercises, at once, only a small influence upon the supply of that commodity; but it starts entrepreneurs guessing whether the higher price will continue. If they decide that it probably will continue, they may start upon the production of a considerable increased supply for a furture date. This decision will affect their current demand for factors; the current position in the factor markets will thus be governed by the way entrepreneurs interpret the rise in the price of the product.

Similarly, the current supply of a commodity depends not so much upon what the current price is as upon what entrepreneurs have expected it to be in the past. It will be those past expectations, whether right or wrong, which mainly govern current output; the actual current price has a relatively small influence.

This is the first main crux of dynamic theory; and it marks the first parting of the ways. Either we have to face up to the difficulty, and allow deliberately for the fact that supplies (and ultimately demands, too) are governed by expected prices quite as much as by current prices; or we have to evade the issue by concentrating on the case where these difficulties are at a minimum. (1946:116–17)

The method of 'evading' the issue, Hicks labels Austrian, with its 'concentration on the case of a Stationary State', the very method used in the *Theory of Wages*.

The criticisms of stationary state theory are pointed. In *Value and Capital*:

Although it is my firm belief that the stationary state is, in the end, nothing but an evasion, nevertheless it has played so large a part in modern economic thought that we must give it some attention. The stationary state is that special case of a dynamic system where tastes, technique, and resources remain constant through time. We can reasonably assume that experience of these constant conditions will lead entrepreneurs to expect their continuance; so that it is not necessary to distinguish between price-expectations and current prices, for they are all the same. We can assume, too, that entrepreneurs did expect in the past that today's prices would be what they now turn out to be; so that the supplies of commodities are fully adjusted to their prices. (1946:117)

As such, the stationary state is:

quite incompetent to deal properly with capital, or interest, or trade fluctuations, or even money—problems where the dating of quantities is of the first importance . . . It is only in the stationary state that actual prices do not need to be distinguished from expected prices; that income does not need to be distinguished from product; that money rates of interest do not need to be distinguished from real rates of interest, and interest rates for one period of lending from interest rates for another. The stationary state has positively impeded the development of the theory of interest, by leaving out so many vital aspects. (1946:116,119)

And, in *Capital and Growth*:

The crucial assumption of static theory (without which it could not have developed as it has been developed) is that a static economy (static, because tastes and resources are unchanging) can be treated as if it were in equilibrium: the quantities produced and consumed will be (near enough) the equilibrium quantities that have just been described. (1965:116)

The problem for Hicks (1946:119) is that 'although it would always be recognized that the actual economy is never in fact stationary, nevertheless stationary state theorists naturally regarded reality as "tending" towards stationariness; though the existence of such a tendency is more than questionable.'[7] The theory 'tells us that if we got to a stationary state, then (other things being equal) we would stick; but it gives us no indication that we are in fact aiming for such a position; for it can tell us nothing about anything actual at all.'

Hicks's (1946:132ff) own contribution in *Value and Capital*, as is well known, comes in two forms: a theory for a 'temporary equilibrium', and one for an 'equilibrium over time'. The former uses the device of a 'week' (as opposed to Marshall's 'market day'), over which market prices are assumed to be constant while individuals form plans, based upon prices expected in the next 'week', for future consumption and production. Static price-level expectations are assumed, so that prices this 'week' are the estimates for prices next 'week'. On each 'Monday', markets clear (as a 'temporary equilibrium'), contracts and plans are made, and resources are used to produce output over the remaining six 'days'.[8]

Since expected and actual prices will generally vary, stringing together a series of 'temporary equilibria' generates, for Hicks, an 'equilibrium over time'. Contrary to the stationary state idea, equilibrium does *not* imply (unless, of course, expected and actual prices happen to be equal) constant relative prices—actual prices may rise or fall through time, with no necessary 'convergence' to stationary values.

As a special case of 'equilibrium over time', Hicks (1946: 136–8) also outlines the properties of a 'Futures Economy'; essentially, although not necessarily, the case of perfect foresight, where the existence of a full set of perfect futures markets gives, in the initial time period, not only current prices, but market-clearing prices for all successive time-periods.[9] Once again, the time-sequence of prices may not be stationary, but if initial prices are properly chosen, the system will 'converge' to a stationary or steady-state equilibrium.[10] A full set of futures prices (i.e., a Neo-Walrasian intertemporal equilibrium), in turn, defines different interest rates, or own-rates of return, as many as there are goods and time-periods in the system.[11]

Nevertheless, even with all of Hicks's criticisms of the stationary state, there is still a direct correspondence between a Neo-Walrasian intertemporal equilibrium and the long-run theory of Walras and Wicksell. The latter is the stationary outcome of the former. For the non-stationary case, a Neo-Walrasian intertemporal equilibrium requires that *nominal* (and not stationary real) rates of return be equalized across various assets.[12]

To see this, in the simplest way, assume the case of perfect futures markets. All goods are distinguished by type, location and their point in time. Market-clearing gives a set of nT constant-value or discounted prices at the initial point in time (say) $t = 1$, for n goods ($j = 1, 2, \ldots, n$) and T 'elementary time-periods', ($t = 1, 2, \ldots, T$). Implicit in the set of equilibrium prices are own-rates of return defined across goods and time-periods. Define p_j^t to be the discounted price of good j in period t. The own-rate of return, for example, for good one, from period 1 to t, is

$$r_1(1, t) = (p_1^1 - p_1^t)/p_1^t \qquad \text{(C.1)}$$

so that r_1 measures how much extra (per-unit) of good one can be obtained in week t, by deferring 'delivery' from week one to t. In general, once again, there will be as many different own-rates (some may even be negative if intertemporal prices rise over time), as there are goods and 'elementary time-periods'.

In short, there will only be *one* own-rate of return—the case of long-run equilibrium, either stationary or in steady-state growth—if relative prices, given the choice of *numéraire*, remain constant over time. More precisely, the price of good two in period one, relative to t, is

$$p_2^1 = [1 + r_2(1, t)]p_2^t \qquad \text{(C.2)}$$

using equation (C.1). Set $p_1^1 \equiv 1$, as the *numéraire*. The value of any p_j^t, in terms of p_1^1, depends, first, on how the present value of good one, in period t, stands in relation to its value in period one, i.e., since

$$p_j^t = [1 + r_1(1, t)]^{-1} \cdot (p_j^t/p_1^t) \qquad \text{(C.3)}$$

the value of p_j^t depends on the time-profile of the *numéraire*, and, second, on how the value of good j stands in period t, relative to good one at t. If own-rates are identical

$$1 + r_2(1, t) = p_2^1/p_2^t = 1 + r_3(1, t) = p_3^1/p_3^t \qquad \text{(C.4)}$$

then

$$p_2^1/p_3^1 = p_2^t/p_3^t \qquad \text{(C.5)}$$

or $r \equiv r_2 = r_3$ for, say, good two relative to three. If equation (C.5) holds for all goods and all time-periods, relative prices are constant throughout; there is, in terms of the *numéraire*, a single own-rate of return or a unique rate of interest.

What if equation (C.5) does not hold? Imagine that production takes place with constant returns to scale and index outputs by q_j, inputs by a_{ij}, for (i = 1, 2, . . . n), and assume that labour, or a_{0j}, is the sole 'original' or non-produced factor of production. Discounted profits are

$$\Pi_t = p_j^{t+1} \cdot q_j^{t+1} - \sum_{i=0}^{n} p_i^t a_{ij}^t \qquad \text{(C.6)}$$

for prices in constant-value terms, from periods t to t+1. To define current-value prices, essentially the values derived from discounted prices, with the price of the *numéraire* good set equal to one in *each* period, multiply each constant-value price by

$$\beta_1^t \equiv \beta^t = 1/p_1^t \qquad \text{(C.7)}$$

so that own-rates become

$$r_{jt} \equiv r_j(1, t) = (\beta^{t+1}/\beta^t) \cdot (P_j^t/P_j^{t+1}) \qquad \text{(C.8)}$$

for P_j^t the current-value or undiscounted price of j at time t. For the case of the *numéraire*

$$r_t = (\beta^{t+1} - \beta^t)/\beta^t \tag{C.9}$$

for $r_t \equiv r_1(t, t + 1)$.

The expression for undiscounted profits is simply

$$\bar{\Pi}_t = P_j^{t+1} \cdot q_j^{t+1} - \sum_{i=0}^{n} P_i^t a_{ij}^t - r_t \sum_{i=0}^{n} P_i^t a_{ij}^t \tag{C.10}$$

as an analog to equation (C.6). Note that transferring (C.10), using equation (C.9), back to constant-value prices

$$\Pi_t = \beta^{t+1} \cdot p_j^{t+1} \cdot q_j^{t+1} - \beta^t \sum_{i=0}^{n} p_i^t a_{ij}^t - (\beta^{t+1}/\beta^t - 1)\beta^t \sum_{i=0}^{n} p_i^t a_{ij}^t \tag{C.11}$$

returns (C.6). Simplifying equation (C.10) gives

$$\bar{\Pi}_t = P_j^{t+1} \cdot q_j^{t+1} - (1 + r_t) \sum_{i=0}^{n} P_i^t a_{ij}^t \tag{C.12}$$

with constant returns to scale and competition resulting in zero economic profits. In terms of the value of net output, equation (C.12) becomes

$$r_t = \left(P_j^{t+1} \cdot q_j^{t+1} - \sum_{i=0}^{n} P_i^{t+1} \cdot a_{ij}^t\right) \Big/ \sum_{i=0}^{n} P_i^t a_{ij}^t + (P_i^{t+1} - P_i^t)/P_i^t \tag{C.13}$$

as a measure of the rate of return r_t.[13] Note that the last expression in (C.13) is simply

$$(P_i^{t+1} \cdot a_{ij}^t - P_i^t a_{ij}^t)/P_i^t a_{ij}^t \tag{C.14}$$

in terms of the total value of inputs.

Now, once again, if the system is stationary or in steady-state growth, there is no distinction between relative prices at different points of time. Assume, also, that labour is paid *post factum*, or after the period of production, and express price equations, in terms of per-unit output, for per-unit input coefficients (say) $\mathbf{a}_{ij}$. Equation (C.12) becomes

$$P_j = (1 + r) \sum_{i=1}^{n} P_i \mathbf{a}_{ij} + W\mathbf{a}_{0j} \tag{C.15}$$

for W the wage rate. Although derived from a Neo-Walrasian framework, this indeed is the same equation given in Sraffa (1960). It is also identical to Wicksell's price equations (see Section 4.3c above) ignoring land, in terms of 'dated' labour inputs, and to those of Walras, from Part V of the *Elements* (see Section 3.4a above). These systems are all directly comparable. Note, as well, assuming two-outputs and two-produced inputs, physically identical to q_j, that (C.15) represents two independent equations in three unknowns: P_2, W and r, with $P_1 \equiv 1$ as *numéraire*, i.e., there is a 'missing equation'.[14]

If the economy is not stationary, equation (C.13) holds. The numerator, however, for the first term in (C.13)

$$R_i^{t+1} \equiv P_j^{t+1} - \sum_{i=1}^{n} P_i^{t+1} \cdot \mathbf{a}_{ij}^{t} \qquad \text{(C.16)}$$

is simply the (per unit) net rental for each input. (With complete substitutability, this reads as the value of the net marginal product.) The denominator is the value of inputs used to produce each q_j or (say) V_i^t. Equation (C.13) becomes

$$r_t = R_i^{t+1}/V_i^t + (V_i^{t+1} - V_i^t)/V_i^t \qquad \text{(C.17)}$$

or, with efficient markets, a portfolio equilibrium condition that specifies that nominal returns across inputs (or any asset) are equal.

Equation (C.17) can also be derived, in an analogous way, from the dual-formulaton of an efficient input-output system, e.g., Solow (1959:44–8). Undiscounted net rentals must satisfy

$$R_i^{t+1}(1 + r_t)^{-t-1} = P_i^t(1 + r_t)^{-t} - P_i^{t+1}(1 + r_t)^{-t-1} \qquad \text{(C.18)}$$

or, rearranging

$$r_t = R_i^{t+1}/P_i^t + (P_i^{t+1} - P_i^t)/P_i^t \qquad \text{(C.19)}$$

returns equation (C.17). In continuous form, (C.19) is

$$r = R_i/P_i + \dot{P}_i/P_i \qquad \text{(C.20)}$$

or nominal returns are the sum of real returns [see equation 3.1.9 above], plus capital gains or losses.[15] The stationary outcome to equation (C.20), as the number of analytic time-periods T gets infinitely large, is simply

$$r = (R_i - \delta_i P_i)/P_i \qquad \text{(C.21)}$$

or a uniform rate of return across all capital goods, for a rate of depreciation δ_i.

But this is the same expression for uniform returns, again, derived from a Neo-Walrasian framework, used by Walras (Part V of the *Elements*) and Wicksell (the wine models, the two-consumption good, two-capital good models, 'full' stationary equilibrium, etc.). There is no mistaking the 'essential continuity', at least as far as the theory of long-run equilibrium is concerned, in the history of economic thought.

Notes to Appendix C

1 Note Hicks (1963 [1932]:20–1): 'Since the whole conception of marginal productivity depends on the variability of industrial methods, little advantage seems to be gained from the attempt which is sometimes made to define a "short period marginal product"—the additional production due to a small increase in the quantity of labour, when not only the quantity, but also the form, of the co-operating capital is supposed unchanged . . . It seems much best to restrict the term "marginal product" to the sense in which we have used it in defining the conditions of full equilibrium.'

2 Hicks (1963:314), however, attributes this kind of theory only to Böhm-Bawerk and Wicksell.

3 Hicks (1946:117) asserts that both Böhm-Bawerk and Wicksell defined the stationary state as an equilibrium conditional on the assumption of zero net savings. In fact, this is only true for Wicksell. As Wicksell is careful to point out, Böhm-Bawerk conducts his analysis solely in terms of a non-stationary economy. See Sections 4.2a and 4.2c above.

4 See Hicks (1963:205–6, 314, 362), and note that Hicks (1946: 116–17) defines the idea of a 'stationary state' by a condition of zero net savings—the same device used in *The Theory of Wages*.

In Garegnani's defense, Hicks does not make the condition of zero net savings sufficiently explicit in *The Theory of Wages*. Only after reading *Value and Capital*, and Hicks's 1963 commentary on *The Theory of Wages*, does his approach to closing the system of equations for a stationary long-run equilibrium become especially clear.

It is also important to recall that the formal model contained in *The Theory of Wages*, as opposed to the informal discussion, with a number of different capital goods, outlined above, runs in terms of a single homogeneous capital good; Hicks (1963:338): 'The formal model which I had in mind consisted of a (static) system with only one (homogeneous) product and two (homogeneous) factors . . . With constant returns to scale, the relative shares of the factors depend only on the ratio of the quantities of the factors.'

5 Cf. Garegnani (1976:31–3) and Petri (1978:255) who argue just the opposite.

6 See Milgate (1979) who provides evidence for the importance of Lindahl (1970 [1939]), in addition to Hayek (1984 [1928]), on Hicks's 'conversion'.

7 In Hicks (1979:57), the criticism of the stationary state turns on its inability to define a temporal sequence and problems with capital reswitching and reversal: 'It is nevertheless apparent, when the issue is posed in this way, that it is quite hopeless to string such equilibria together, making them successive in time, so as to form a growth model. No means is shown, no means can be shown, by which an economy can pass from one such equilibrium to another. So nothing can be shown about the *accumulation* of capital . . . Where the method belongs is in comparative statics; but even here it has a weakness, which cannot be denied . . . it requires us to be able to say that when two static states are being compared, the volume of capital in the one *is* greater than in the other; it requires us, at least to that extent, to be able to measure capital.'

8 Hicks (1946:128–9), like Wicksell, is well aware that such equilibria are generally path-dependent. To avoid the problem, however, he assumes that trades at 'false prices', if they occur, result in largely offsetting income effects, so that the 'market will finish up very close to the equilibrium price.' Recall, that Wicksell [see fn. 21, Chapter 2 above] ignores the 'transition path' by simply assuming that the 'final price equilibrium is established from the very beginning.'

In this sense, Walras [see Chapter 2 above] commits two errors. First, there is no recognition whatsoever of the problem of path-dependency. When Walras does take into account the presence of disequilibrium transactions he simply asserts that the same 'true' equilibrium prices would be obtained, with or without trades at false prices. Second, although Walras correctly argues that prices 'cried' in terms of the *numéraire* rule out arbitrage, he also mistakenly claims that the resulting equilibrium must be unique and therefore stable.

9 Hicks (1963:307) does not find the construction theoretically appealing: 'There is indeed no particular difficulty about constructing a Paretian model of a dynamic economy (in which the 'data' are changing over time); but since the prices relevant to present decisions would be future prices as well as present prices, the dynamic model could only be made consistent if the prices expected (to rule at time t) at all earlier dates were the same, and the same as those which actually equated supplies and demands at time t. Dynamic equilibrium was an equilibrium of perfect foresight. This was obviously unrealistic, even more unrealistic than the static scheme which I had used in *The Theory of Wages*.'

10 See Appendix D below.

11 The principal sources for Neo-Walrasian theory are Debreu (1959), Solow (1959) and Malinvaud (1953). Bliss (1975), Burmeister (1980) and Malinvaud (1972) are useful references.

12 Even here, however, there is an analog to Walras and Wicksell. A non-stationary, intertemporal equilibrium is a theoretically accurate description, over a sequence of time-periods, of the 'move' from Parts IV to V of the *Elements*; a characterization that Walras, without the formal device of difference equations, could not perform. In addition, if the system is restricted to 'circulating' capital goods only, so that capital gains and losses can be ignored, uniform nominal returns (in non-stationary growth) also characterizes Wicksell's sequence of 'preliminary' long-run equilibria to a 'complete' stationary state.

Milgate (1979:1), Garegnani (1976:36–9) and Petri (1978:255–8) err by asserting that there are no conditions under which a uniform rate of return, however defined, can be obtained in an Arrow-Debreu economy. Garegnani goes on to claim that such a system is directly analogous to Part IV of the *Elements*, where there are as many different input-specific rentals as there are 'capital goods proper'. In fact, this case only applies to a Neo-Walrasian 'atemporal equilibrium'. See Malinvaud (1960–61).

13 Bliss (1975:10, 55), Hahn (1975:360) and Hahn (1982) seem to miss the point that there is a uniform *nominal* rate of return in Neo-Walrasian intertemporal prices. Although that rate is clearly contingent on the choice of *numéraire*, once chosen, returns are equalized.

14 Sraffa (1960) takes input coefficients as fixed and given; for a neoclassical, these are the result of firms' cost-minimizing, with all prices and the rate of return as parametric. See fn. 67, Chapter 4 above.

15 In terms of Hicks's 'temporary equilibrium', with static price-level expectations, the expected value of the capital gains/losses term is zero, i.e., there is an expected uniform real return. Cf. the interpretation of Walras [see Chapter 3 above] in terms of static expectations.

Note Hicks (1963:346): 'Heterogeneity of product and heterogeneity of capital stock can be dealt with by using our formal theory of many factors and many products; and it should be noticed that this makes it unnecessary to insist that the economy be in a state of perfect equalization of yields on every sort of capital good.' If Hicks means that *actual* real returns will not be equal (setting aside the case of semi-stationary growth) he is correct.

Appendix D

It is worth noting that the stability or 'convergence' to a stationary equilibrium, defined by equation (C.15) in Appendix C, is a difficult problem. In general, with 'myopic perfect foresight', i.e., all agents know with certainty the right-hand time-derivative of P(t), but not the entire time-profile of P(t), the underlying dynamic system is saddlepoint unstable [e.g., Shell and Stiglitz (1967)]. Except for the 'right' initial value of P(0), the economy will not 'converge' to its stationary outcome.

This is a key point to keep in mind. Duménil and Levy (1985:335), in what is likely to become an influential commentary on Hahn (1982), *mistakenly* argue to the contrary, i.e., they argue that the stability for this kind of stationary equilibrium is straightforward: 'The intertemporal equilibrium model automatically adjusts its prices toward [stationary] prices of production if a sufficient number of periods exist—that is, the intertemporal equilibrium over an infinite horizon generally "converges" toward [Sraffian] prices of production.'

To see the problem, in a simple way, consider the following log-linear, first-order difference equation

$$E_t P_i(t + 1) = \lambda P_i(t) - x(t) \tag{D.1}$$

as an analog to (C.13), for E_t an expectations operator for forecasts at time t for t + 1, λ a positive constant and x(t) a 'forcing variable' or, here, the value of net rentals as a 'market fundamental'. Define

$$\pi(t) \equiv E_t P_i(t + 1) - P_i(t) = (\lambda - 1)P_i(t) - x(t) \tag{D.2}$$

as the expected instantaneous rate of change in $P_i(t)$. With myopic perfect foresight

$$\pi(t) = dP_i/dt \equiv \dot{P}_i \tag{D.3}$$

so that

$$\dot{P}_i + (1/a)\dot{P}_i(t) = -x(t) \tag{D.4}$$

for $1 - (1/a) = \lambda$ and $a < 0$. The complementary solution to equation (D.4) is

$$\dot{P}_i(t) = be^{-t/a} \tag{D.5}$$

for b an arbitrary constant, where

$$d[e^{t/a}P_i(t)]/dt = e^{t/a}[\dot{P}_i + (1/a)P_i(t)]. \tag{D.6}$$

With equation (D.6), multiplying (D.4) through by $e^{t/a}$ and integrating gives the general solution to (D.4)

$$P_i(t) = \int_{s=t}^{\infty} e^{(s-t)/a}x(s)ds + ce^{-t/a} \tag{D.7}$$

for c a constant of integration. If $x(t) = \bar{x}$, i.e., if net rentals are constant over time, (D.7) becomes

$$P_i(t) = -a\bar{x} + [a\bar{x} + P_i(0)]e^{-t/a} \tag{D.8}$$

for $c = P_i(0)$, or the value of prices in the initial time period.

For the case in which $x(t) = \bar{x}$, the 'right' value for c is $-a\bar{x}$, i.e., in this one-dimensional, log-linear case, $P_i(0)$ must equal the market fundamental for all t. In a general Neo-Walrasian system, this implies that real returns are always equalized. There can never be any capital gains or losses; the system must be in long-run equilibrium to begin with. If $P_i(0) \neq -a\bar{x}$, and since $-t/a > 0$, prices continuously diverge from their stationary values in the nature of a 'speculative bubble', albeit a 'bubble' that never bursts. (Since agents know with certainty that $\pi(t) > 0$, they invest in this particular capital good and indeed drive its price higher.) If rentals are not constant over time, the limit of the expression in the integral in (D.7), as $s \to \infty$, must also converge.

The bottom line is straightforward. Generally speaking, Neo-Walrasian prices will not converge to long-run (stationary) prices of production. There is nothing inherent in the competitive mechanism to guarantee the 'right' initial value for $P_i(0)$. The presence of an intertemporal auctioneer, alternatively, simply imposes stability, i.e., it imposes a stationary outcome, in terms of equation (C.13), on an otherwise unstable process, by not only picking the 'right' initial value for prices, but by making certain that no price becomes zero or infinite at some time t less than infinity. This guarantees—the case of 'complete' perfect foresight—that $P_i(t)$ will converge. But it also necessarily implies that all future prices are known in the initial time period.[1]

Note to Appendix D

1 For an application of these results to macroeconomics, see Kompas and Spotton (1989).

References

Arrow, K. J., and Hahn, F. H. (1971), *General Competitive Analysis*, Amsterdam, North-Holland.

Arrow, K. J., and Hurwicz, L. (1958), 'On the Stability of Competitive Equilibrium, I', *Econometrica*, 26, 522–52.

Bertrand, Joseph (1883), 'Théorie Mathématique de la Richesse Sociale', *Bulletin des Sciences Mathématiques et Astronomiques*, 293–303, in Cournot, Augustin (1938), *Recherches sur les Principes Mathématiques de la Théorie des Richesses*, Paris, Marcel Rivière, 233–50.

Bliss, Christopher J. (1975), *Capital Theory and the Distribution of Income*, Amsterdam, North-Holland.

Böhm-Bawerk, Eugen von (1889), *The Positive Theory of Capital*, Vol. II of *Capital and Interest*, George D. Huncke, trans. (1959), South Holland, Ill., Libertarian press.

Bortkiewicz, Ladislaus von (1890), 'Review of *Eléments d'Économie Politique Pure, ou Théorie de la Richesse Sociale*', 2nd ed., by Léon Walras, *Review d'Économie Politique*, 4, 80–6.

Burmeister, Edwin (1980), *Capital Theory and Dynamics*, Cambridge, Cambridge University Press.

Cass, David (1973), 'On the Wicksellian Point-Input Point-Output Model of Capital Accumulation: A Modern View (or Neoclassicism Slightly Vindicated)', *Journal of Political Economy*, 81, 71–97.

Clark, John Bates (1965 [1899]), *The Distribution of Wealth: A Theory of Wages, Interest and Profits*, New York, Kelley.

Collard, D. (1973), 'Léon Walras and the Cambridge Caricature', *Economic Journal*, 83, 465–75.

Cournot, Augustin (1838), *Researches into the Mathematical Principles of the Theory of Wealth*, Bacon, N. T., trans. (1960), London, Hafner.

Debreu, G. (1959), *Theory of Value: An Axiomatic Analysis of Economic Equilibrium*, New Haven, Conn., Yale University Press.

Diewert, W. E. (1977), 'Walras's Theory of Capital Formation and the Existence of a Temporary Equilibrium', in *Equilibrium and Disequilibrium in Economic Theory*, G. Schwödiauer, ed., Dordrecht, Holland, Riedel, 73–126.

Dorfman, Robert (1959a), 'A Graphical Exposition of Böhm-Bawerk's Interest Theory', *Review of Economic Studies*, 26, 153–8.

Dorfman, Robert (1959b), 'Waiting and the Period of Production', *Quarterly Journal of Economics*, 73, 351–72.

Duménil, G., and Levy, D. (1985), 'The Classicals and the Neoclassicals: A Rejoinder to Frank Hahn', *Cambridge Journal of Economics*, 9, 327–45.

Edgeworth, F. Y. (1967 [1881]), *Mathematical Psychics: An Essay on the Application of Mathematics to the Moral Sciences*, New York, Kelley.

Edgeworth, F. Y. (1889), 'On the Application of Mathematics to Political Economy', in *Papers Relating to Political Economy*, Vol. II (1925), New York, Burt Franklin, 273–312.

Edgeworth, F. Y. (1891), 'La Théorie Mathématique de l'Offre et de la Demande et le Coût de Production', *Review d'Économie Politique*, 5, 10–28.

Edgeworth, F. Y. (1904), 'The Theory of Distribution', *Quarterly Journal of Economics*, 18, 159–219, in *Papers Relating to Political Economy*, Vol. I (1925), New York, Burt Franklin, 13–59.

Fisher, Franklin M. (1983), *Disequilibrium Foundations of Equilibrium Economics*, Cambridge, Cambridge University Press.

Garegnani, Pierangelo (1976), 'On a Change in the Notion of Equilibrium in Recent Work on Value and Distribution', in *Essays in Modern Capital Theory*, Brown, M., Sato, K., and Zarembka, P., eds., Amsterdam, North-Holland, 25–45.

Gifford, C. (1935), 'The Period of Production under Continuous Input and Output in an Unprogressive Community', *Econometrica*, 3, 199–212.

Hahn, F. H. (1975), 'Revival of Political Economy: The Wrong Issues and the Wrong Argument', *Economic Record*, 51, 360–4.

Hahn, F. H. (1982), 'The Neo-Ricardians', *Cambridge Journal of Economics*, 6, 353–74.

Harcourt, G. C. (1972), *Some Cambridge Controversies in the Theory of Capital*, Cambridge, Cambridge University Press.

Hayek, F. A. (1928), 'Intertemporal Price Equilibrium and Movements in the Value of Money', in *Money, Capital and Fluctuations: Early Essays* (1984), Chicago, University of Chicago Press, 71–117.

Hicks, J. R. (1963 [1932]), *The Theory of Wages*, 2nd ed., London, Macmillan.

Hicks, J. R. (1934), 'Léon Walras', *Econometrica*, 2, 338–48.

Hicks, J. R. (1935), 'Wages and Interest: The Dynamic Problem', *Economic Journal*, 45, 456–68, in *Money, Interest and Wages: Collected Essays on Economic Theory*, Vol. II (1982), Cambridge, Harvard University Press, 64–79.

Hicks, J. R. (1946), *Value and Capital*, 2nd ed., Oxford, Oxford University Press.

Hicks, J. R. (1956), *A Revision of Demand Theory*, Oxford, Oxford University Press.

Hicks, J. R. (1963), 'Commentary on *The Theory of Wages*', in *The Theory of Wages*, 2nd ed., London, Macmillan, 305–72.

Hicks, J. R. (1965), *Capital and Crowth*, Oxford, Oxford University Press.

Hicks, J. R. (1973), *Capital and Time: A Neo-Austrian Theory*, Oxford, Oxford University Press.

Hicks, J. R. (1979), *Causality in Economics*, Oxford, Basil Blackwell.

Hirshleifer, J. (1967), 'A Note on the Böhm-Bawerk/Wicksell Theory of Interest', *Review of Economic Studies*, 34, 191–9.

Hirshleifer, J. (1970), *Investment, Interest and Capital*, Englewood Cliffs, N. J., Prentice-Hall.

Hollander, Samuel (1979), *The Economics of David Ricardo*, Toronto, University of Toronto Press.

Hollander, Samuel (1982), 'On The Substantive Identity of the Ricardian and Neoclassical Conceptions of Economic Organization: The French Connection in British Classicism', *Canadian Journal of Economics*, 15, 586–612.

Hollander, Samuel (1987), *Classical Economics*, Oxford, Basil Blackwell.

Howard, M. C. (1980), 'Austrian Capital Theory: An Evaluation in Terms of Piero Sraffa's Production of Commodities by Means of Commodities', *Metroeconomica*, 32, 1–23.

Jaffé, William (1935), 'Unpublished Papers and Letters of Léon Walras', *Journal of Political Economy*, 43, 187–207, in *William Jaffé's Essays on Walras*, Donald A. Walker, ed. (1983), Cambridge, Cambridge University Press, 17–35.

Jaffé, William (1942), 'Léon Walras' Theory of Capital Formation,' in *Studies in Mathematical Economics and Econometrics*, Oscar Lange, ed., Chicago, University of Chicago Press, 37–48.

Jaffé, William (1964), 'New Light on an Old Quarrel: Barone's Unpublished Review of Wicksteed's "Essay on the Coordination of the Laws of Distribution" and Related Documents,' *Cahiers Vilfredo Pareto*, 3, 61–102, in *William Jaffé's Essays on Walras*, Donald A. Walker, ed. (1983), Cambridge, Cambridge University Press, 176–212.

Jaffé, William (1965), *Correspondence of Léon Walras and Related Papers*, 3 vols., Amsterdam, North-Holland.

Jaffé, William (1967), 'Walras' Theory of *Tâtonnement*: A Critique of Recent Interpretations', *Journal of Political Economy*, 75, 1–19, in *William Jaffé's Essays on Walras*, Donald A. Walker, ed. (1983), Cambridge, Cambridge University Press, 221–43.

Jaffé, William (1980), 'Walras' Economics as Others See It', *Journal of Economic Literature*, 18, 528–49.

Jaffé, William (1981), 'Another Look at Walras' Theory of *Tâtonnement*', *History of Political Economy*, 13, 313–36, in *William Jaffé's Essays on Walras*, Donald A. Walker, ed. (1983), Cambridge, Cambridge University Press, 244–66.

Jevons, William Stanley (1965 [1911]), *The Theory of Political Economy*, 4th ed., New York, Kelley.

Kaldor, N. (1934), 'A Classificatory Note on the Determinateness of Equilibrium', *Review of Economic Studies*, 1, 122–36, in *Essays on Value and Distribution*, (1960), Glencoe, Ill., Free Press, 13–33.

Kennedy, Charles (1968), 'Time, Interest and the Production Function', in *Value, Capital and Growth: Papers in Honour of Sir John Hicks*, J. N. Wolfe, ed., Edinburgh, Edinburgh University Press, 275–89.

Kompas, Tom, and Spotton, Brenda (1989), 'A Note on "Rational" Speculative Bubbles', *Economics Letters*, 30, 327–31.

Kuenne, Robert E. (1962), 'The Stationary State and the Technological Superiority of Present Goods', *Quarterly Journal of Economics*, 76, 648–52.

Kuenne, Robert E. (1963), *The Theory of General Economic Equilibrium*, Princeton, Princeton University Press.

Kuenne, Robert E. (1971), *Eugen von Böhm-Bawerk*, New York, Columbia University Press.

Lange, Oskar (1936), 'The Place of Interest in the Theory of Production', *Review of Economic Studies*, 3, 159–92.

Lindahl, Erik. (1970 [1939]), *Studies in the Theory of Money and Capital*, New York, Kelley.

Lutz, Friedrich A. (1968), *The Theory of Interest*, Chicago, Aldine.

Malinvaud, E. (1953), 'Capital Accumulation and the Efficient Allocation of Resources', *Econometrica*, 21, 233–68.

Malinvaud, E. (1960–61), 'The Analogy Between Atemporal and Intertemporal Theories of Resource Allocation', *Review of Economic Studies*, 28, 143–60.

Malinvaud, E. (1972), *Lectures on Microeconomic Theory*, Amsterdam, North-Holland.

Marshall, Alfred. (1930 [1920]), *Principles of Economics*, 8th ed., London, Macmillan.

Metzler, L. (1945), 'Stability of Multiple Markets: The Hicks Conditions', *Econometrica*, 13, 277–92.

Milgate, Murray (1979), 'On the Origin of "Intertemporal Equilibrium"', *Economica*, 46, 1–10.

Mill, John Stuart (1871), *Principles of Political Economy with Some of Their Applications to Social Philosophy*, 7th ed., in the *Collected Works of John Stuart Mill*, Vol. III, J. Robson, ed. (1965), Toronto, University of Toronto Press.

Mirowski, Philip (1989), *More Heat than Light: Economics as Social Physics, Physics as Nature's Economics*, Cambridge, Cambridge University Press.

Montgomery, William David (1971), 'An Interpretation of Walras' Theory of Capital as a Model of Economic Growth,' *History of Political Economy*, 3, 278–97.

Morishima, Michio (1964), *Equilibrium, Stability and Growth: A Multi-Sectoral Analysis*, Oxford, Oxford University Press.

Morishima, Michio (1977), *Walras' Economics: A Pure Theory of Capital and Growth*, Cambridge, Cambridge University Press.

Morishima, Michio (1980), 'W. Jaffé on Léon Walras: A Comment', *Journal of Economic Literature*, 18, 550–58.

Negishi, Takashi (1962), 'The Stability of a Competitive Economy: A Survey Article', *Econometrica*, 30, 635–69.

Negishi, Takashi (1982a), 'Wicksell's Missing Equation: A Comment', *History of Political Economy*, 14, 310–11.

Negishi, Takashi (1982b), 'Wicksell's Missing Equation and Böhm-Bawerk's Three Causes of Interest in a Stationary State', *Zeitschrift für Nationalökonomie*, 42, 161–74.

Negishi, Takashi (1989), *History of Economic Theory*, Amsterdam, North-Holland.

Neuberger, E. (1960), 'Waiting and the Period of Production', *Quarterly Journal of Economics*, 74, 150–3.

Newman, Peter (1965), *The Theory of Exchange*, Englewood Cliffs, N. J., Prentice-Hall.

Pareto, Vilfredo (1964 [1896–7]), *Cours d'Économie Politique*, 2 vols., Geneva, Librairie Droz.

Pasinetti, Luigi L. (1977), *Lectures on the Theory of Production*, New York, Columbia University Press.

Patinkin, Don (1965), *Money, Interest and Prices: An Integration of Monetary and Value Theory*, 2nd ed., New York, Harper and Row.

Petri, Fabio (1978), 'The Difference Between Long-Period and Short-Period General Equilibrium and the Capital Theory Controversy', *Australian Economic Papers*, 17, 246–60.

Robinson, Joan (1961), 'Prelude to a Critique of Economic Theory', *Oxford Economic Papers*, 13, 52–8.

Robinson, Joan (1969), *The Accumulation of Capital*, 3rd ed., London, Macmillan.

Samuelson, Larry (1982a), 'On Wicksell's Missing Equation', *History of Political Economy*, 14, 301–7.

Samuelson, Larry (1982b), 'The Wicksell Effect in a Growing Economy', *History of Political Economy*, 14, 447–60.

Samuelson, Paul A. (1943), 'Dynamics, Statics and the Stationary State', *The Review of Economics and Statistics*, 25, 194–214.

Sandelin, Bo (1980), 'Wicksell's Missing Equation: The Production Function and the Wicksell Effect', *History of Political Economy*, 12, 29–40.

Sandelin, Bo (1982), 'On Wicksell's Missing Equation: A Comment', *History of Political Economy*, 14, 308–9.

Schumpeter, Joseph A. (1961 [1934]), *The Theory of Economic Development: An Inquiry into Profits, Capital, Credit, Interest, and the Business Cycle*, Redvers Opie, trans., London, Oxford University Press.

Schumpeter, Joseph A. (1954), *History of Economic Analysis*, New York, Oxford University Press.

Shell, Karl, and Stiglitz, Joseph E. (1967), 'The Allocation of Investment in a Dynamic Economy', *Quarterly Journal of Economics*, 81, 592–609.

Solow, Robert M. (1959), 'Competitive Valuation in a Dynamic Input-Output System', *Econometrica*, 27, 30–53.

Sraffa, Piero (1960), *Production of Commodities by Means of Commodities: Prelude to a Critique of Economic Theory*, Cambridge, Cambridge University Press.

Stigler, George J. (1961), *Production and Distribution Theories*, London, Macmillan.

Swan, T. W. (1956), 'Economic Growth and Capital Accumulation', *Economic Record*, 32, 334–61.

Uzawa, H. (1960), 'Walras' *Tâtonnement* in the Theory of Exchange', *Review of Economic Studies*, 27, 182–94.

Walker, Donald A. (1970), 'Léon Walras in the Light of his Correspondence and Related Papers', *Journal of Political Economy*, 78, 685–701.

Walker, Donald A. (1972), 'Competitive *Tâtonnement* Exchange Markets', *Kyklos*, 25, 345–63.

Walker, Donald A. (1987a), 'Walras's Theories of Tatonnement', *Journal of Political Economy*, 95, 758–74.

Walker, Donald A. (1987b), 'Edgeworth Versus Walras on the Theory of Tatonnement', *Eastern Economic Journal*, 13, 155–65.

Walras, Léon (1880), 'La Bourse—la Speculation et l'Arbitrage', *Bibliothéque Universelle et Revue Suisse*, 5, 452–72; 6, 66–94.

Walras, Léon (1883), *Théorie Mathématique de la Richesse Sociale*, Lausanne, Corbaz.

Walras, Léon (1936 [1898]), *Etudes d'Économie Sociale* (*Théorie de la Répartition de la Richesse Sociale*), 2nd ed., G. Leduc, ed., Paris, F. Rouge.

Walras, Léon (1954 [1900]), *Elements of Pure Economics*, William Jaffé, trans., 4th ed. def. (1926), London, George Allen and Unwin.

Weizsäcker, C. C. von (1971), *Steady State Capital Theory*, Berlin, Springer-Verlag.

Wicksell, Knut (1954 [1893]), *Value, Capital and Rent*, S. H. Frowein, trans., London, George Allen and Unwin.

Wicksell, Knut (1897), 'Vilfredo Pareto's *Cours D'Économie Politique*', in *Selected Papers on Economic Theory*, Reginald S. Stedman, trans. (1958), London, George Allen and Unwin, 141–51.

Wicksell, Knut (1965 [1898]), *Interest and Prices: A Study of the Causes Regulating the Value of Money*, R. F. Kahn, trans., New York, Kelley.

Wicksell, Knut (1958 [1900]), 'Marginal Productivity as the Basis of Distribution in Economics', in *Selected Papers on Economic Theory*, Reginald S. Stedman, trans., London, George Allen and Unwin. 93–120.

Wicksell, Knut (1958 [1909]), 'Vilfredo Pareto's *Manuel D'Économie*

Politique', in *Selected Papers on Economic Theory*, Reginald S. Stedman, trans., London, George Allen and Unwin, 159–75.

Wicksell, Knut (1958 [1911]), 'Böhm-Bawerk's Theory of Capital', in *Selected Papers on Economic Theory*, Reginald S. Stedman, trans., London, George Allen and Unwin, 176–85.

Wicksell, Knut (1934 [1911]), *Lectures on Political Economy*, 2nd ed., Vol. I: *General Theory*, E. Classen, trans., London, George Routledge.

Wicksell, Knut (1935 [1911]), *Lectures on Political Economy*, 2nd ed., Vol. II. *Money*, E. Classen, trans., London, George Routledge.

Wicksell, Knut (1934 [1919]), 'Professor Cassel's System of Economics', Solomon Adler, trans., in *Lectures on Political Economy*, Vol. I: *General Theory*, London, George Routledge, 219–57.

Wicksteed, Philip H. (1946 [1933]), *The Common Sense of Political Economy*, Vol. I, London, George Routledge.

Index